UNRAVELING THE UNIVERSE'S MYSTERIES

Explore sciences' most baffling mysteries, including the Big Bang's origin, time travel, dark energy, humankind's fate, and more.

LOUIS A. DEL MONTE

Unraveling the Universe's Mysteries

First edition

International Standard Book Numbers

Paperback: 978-0-9881718-0-0

Hardcover: 978-0-9881718-2-4

eBook: 978-0-9881718-1-7

Publisher: Louis A. Del Monte

Printed in the United States of America

Library of Congress Cataloging

Unraveling the universe's mysteries

(ISBN 9780988171800)

Library of Congress Control Number: 2012915295

Dedication

I would like to dedicate this work to my wife, Diane, the love of my life, and the foundation of our family.

I also dedicate this work to our sons, Bryan and Christian, who are extraordinary fathers, husbands, and men.

Contents

Acknowledgments

I would like acknowledge the following individuals for their assistance in developing and promoting this work.

- Diane Del Monte, Chairwoman of the Board of Directors, Del Monte and Associates, Inc., and Art Director, Del Monte Agency, for her cover art direction and unwavering support of this work
- Nick McGuinness, a dear friend, who generously devoted his time to provide chapter-by-chapter suggestions to clarify the subject matter, and enhance its readability
- Christian Del Monte, Chief Technologist, Del Monte Agency, for his strategic planning and operational management of his agency's promotion of this work
- Bryan Del Monte, President, Del Monte Agency, for providing the full support of Del Monte Agency for cover design, Website development, social media marketing, Internet/traditional advertising, and public relations to promote this work
- Janet Lencowski, Marketing Communications, Del Monte Agency, for assistance with developing the promotional plan and its implementation
- Anthony Hickl, Ph.D., for stimulating conversations on the subject matter and suggestions for improving readability

- Connie Anderson, for editing the book and providing suggestions for improving readability
- David Koepplin, Studio Developer, Del Monte Agency, for book cover design and layout
- Alan Pranke, for the layout of this book and cover consulting

Introduction

The Universe's Unsolved Mysteries

The strides that science made in the Twentieth and early part of the Twenty-First Century are astounding. At the beginning of the Twentieth Century, science held three theories as universal truths, namely:

1) Time was an absolute, independent of distance and movement of observers relative to an event.
2) The universe consisted of the Milky Way galaxy.
3) The universe was eternal and static.

However, the strongly held theories of the greatest scientific minds of the time proved to be false. I will briefly examine each theory and the empirical evidence that caused its demise.

First, the science community up to the early part of the Twentieth Century believed that time was an absolute. This meant time was independent of the position and movement of an observer relative to an event. This almost self-evident theory about time was about to be shattered. In 1905, a young Albert Einstein developed his special theory of relativity. It is termed "special" because it applied only to inertial frames of reference. An inertial frame of reference is one that is at either rest or moving with a constant velocity.

The special theory of relativity offered two hypotheses. 1) The laws of physics are the same in all inertial frames of reference. 2) The speed of light is a constant in a vacuum—independent of the movement of the emission source in all inertial frames. To understand the second hypothesis, consider this example. If you are in an open-top convertible car that is traveling down the highway at sixty miles per hour, you are in an inertial frame of reference. If you throw a ball in the same direction that the car is going, the total speed of the ball will be equal to the speed of the car plus the speed of ball as it leaves your hand. If you are able to throw the ball at thirty miles per hour, the total speed of the ball as it leaves your hand is ninety miles per hour. We get this speed by adding the speed of the car to the speed you are able to throw the ball. Now, let's pretend you have a flashlight, an emission source, and an observer is able to measure the speed of light as it leaves the flashlight. The result the observer would measure is that the speed of light would independent of the car's speed. In effect, the speed of the car does not make the light go faster. Even if the car stops, the speed of light from the flashlight would equal the speed of light of the moving car. For this example, I have ignored atmospheric effects and considered the observer stationary. This is counter intuitive, but true. The speed of light is the same regardless of the speed of the car (inertial frame). The implications of special relativity became enormous. One significant implication demonstrated that time was highly dependent on the relative motion of both the observer and the event. This discovery eventually led to the development of space-time as a coordinate system. The special theory of relativity and the general theory of relativity, two highly successful theories of modern science, use space-time as a coordinate system.

A second theory that the science community held about the universe related to its size. Until the 1917 completion of the 100-inch Hooker Telescope at the Mount Wilson Observatory, science had no way of knowing other galaxies existed. Therefore, the scientific community held that the universe consisted of the Milky Way galaxy, and nothing else. In fact, this is what they taught our grandparents as schoolchildren.

Surprisingly, the German philosopher Immanuel Kant (1724-1804), using reasoning, suggested a hundred years earlier that our galaxy was one of numerous "island universes." Unfortunately, Kant's view would have to wait more than a hundred years for telescope technology to prove him right. Even when early astronomers saw the faint lights of other galaxies in their crude telescopes, they believed the observed phenomena to be part of the Milky Way. That view of the universe was about to dramatically change.

In 1919, a young astronomer, Edwin Hubble, arrived at the Mount Wilson Observatory in California. As chance would have it, his arrival coincided with the completion of the Hooker Telescope. At the time, it was the world's largest telescope, and the only one able to observe other galaxies beyond the Milky Way. In 1924, Edwin Hubble, using the 100-inch telescope at Mt. Wilson, discovered the Andromeda galaxy, a sister galaxy similar to our own Milky Way. This completely shattered another strongly held scientific belief. The universe was larger than previously thought. In fact, today we know that the universe has billions of galaxies.

Lastly, science held that the universe was eternal and static. This meant it had no beginning. Nor would it ever end. In other words, the universe was in "steady state." At the beginning of the Twentieth Century, as I mentioned above, telescopes were crude and unable to focus on other galaxies. In addition, no theories of the universe were causing science to doubt the current dogma of a steady-state universe. All of that was about to change.

In 1916, Albert Einstein developed his general theory of relativity. It was termed "general" because it applied to all frames of reference, not only frames at rest or moving at a constant velocity (inertial frames). The general theory of relativity predicted that the universe was either expanding or contracting. This should have been a pivotal clue that the current scientific view of the universe as eternal and static might be wrong. However, Einstein's paradigm of an eternal and static universe was so strong, he disregarded his own results. He quickly reformulated the equations incorporating a "cosmological constant." With this new mathematical expression plugged into the equations, the equations of general relativity yielded the answer Einstein believed was right. The universe was in a steady state. This means it was neither expanding nor contracting. The world of science accepted this, and continued entrenched in its belief of a steady-state universe. However, as telescopes began to improve, this scientific theory was destined to be shattered.

In 1929, Edwin Hubble, using the new Mt. Wilson 100-inch telescope, discovered the universe was expanding. In time, other astronomers confirmed Hubble's discovery. This forced Einstein to call the cosmological constant his "greatest blunder." This completely shattered the steady-state theory of the universe. In fact, this discovery was going to pave the way to an even greater discovery, the Big Bang theory, but more about that later.

In 1900, and for centuries before that, the greatest scientific minds of the time held the above three theories sacred. However, each theory crumbled as measurement techniques improved, and new theories evolved. This is a pivotal point. Science is rapidly evolving, and scientific knowledge doubles about every 10 years. We are constantly gathering new data that challenges our understanding of science, and that often leads to new mysteries. As soon as we become confident and comfortable in our grasp of reality, a new discovery turns our world upside down. For example, until 1998, every cosmologist knew the universe was expanding, but commonly held the belief that gravity would eventually slow down the expansion, and cause the universe to contract in a "Big Crunch." The Big Crunch would result in an infinitely dense energy point similar to the infinitely dense energy point that existed at the instant before the Big Bang. In effect, the commonly held view was the universe would first expand, via the Big Bang, and then gravity would eventually cause it to contract, via the Big Crunch, to the infinitely dense energy point just prior to the expansion. Their confidence in this view abounded, and three scientists, Saul Perlmutter,

Brian P. Schmidt, and Adam G. Riess, decided to measure it. To the scientific world's astonishment, they discovered the universe was not only expanding, but the expansion was accelerating. In 2011, these three received the Nobel Prize for this remarkable discovery.

The Twentieth Century stands as the golden age of science, yielding more scientific breakthroughs than any previous century. Yet, in the wake of all the scientific breakthroughs over the last century, profound mysteries emerged. To my eye, there appears a direct correlation between scientific discoveries and scientific mysteries. Often, it appears that every significant scientific breakthrough results in an equally profound mystery. I have termed this irony of scientific discovery the Del Monte Paradox, namely:

Each significant scientific discovery results in at least one profound scientific mystery.

I'll use two examples to illustrate this paradox. For our first example, consider the discovery of the Big Bang theory. We will discuss the Big Bang theory in later chapters. For this discussion, please view it as a scientific framework of how the universe evolved. While the scientific community generally accepts the Big Bang theory, it is widely acknowledged that it does not explain the origin of the energy that was required to create the universe. Therefore, the discovery of the Big Bang theory left science with a profound mystery. Where did the energy originate to create a Big Bang? This is arguably the greatest mystery in science, and currently an area of high scientific focus. For the second example, consider the discovery we discussed above—the universe's expansion is accelerating. This leaves us with another profound mystery. What is causing the universe's expansion to accelerate? Numerous theories float within the scientific community to explain these mysteries. None has scientific consensus.

This book will investigate and provide insight on some of science's greatest mysteries. Although there are numerous scientific mysteries, we will concentrate on three main "classes" of mysteries by section:

Section I: What Caused the Big Bang?

Section II: What Mysteries Still Baffle Modern Science?

Section III: Are We Alone?

All are highly active areas of scientific research, and bring us to the edge of scientific knowledge. All influence the direction scientific research is taking. One scientific breakthrough on any one of these mysteries could literally change the world of science.

The scientific community is not in complete consensus with numerous theories forwarded to address the mysteries. This is how it should be, since the theories reside on the edge of scientific knowledge. In a way, this is a righteous thing. Science moves forward via rigorous debate, experimentation, and independent validation of scientific findings and theories. All significant scientific theories have gone through this process. This is the scientific method. Remember that Einstein's special theory of relativity, published in 1905, took about 15 years to gain acceptance by the majority of the scientific community (circa 1920). Here I'll dispel a commonly held belief about Einstein. Most people have heard of Albert Einstein. They consider him one of the greatest scientists that ever lived. They believe that he jotted down equations, and created new theories, while working separate from the rest of the scientific community. This view of Einstein quietly working at his desk and dreaming up theories and equations is completely erroneous. Nothing could be further from the truth. Einstein let the experiments and observations of the scientific community guide his theoretical work. He cared deeply about the acceptance of his theories. In fact, in 1919, three years after publishing his general theory of relativity, he stated, "By an application of the theory of relativity to the taste of readers, today in Germany I am called a German man of science, and in England I am represented as a Swiss Jew. If I come to be regarded as a bête noire (black beast or a person strongly detested) the descriptions will be reversed, and I shall become a Swiss Jew for the Germans and a German man of science for the English!"

Einstein can rest in peace. Science holds the special theory of relativity as the golden standard, having withstood the rigor of over 100 years of scientific investigation. Elements of the general theory of relativity have also withstood vigorous investigation. To that point, scientists believe that other theories, such as string theory and dark energy, which we discuss in later chapters, needs to meet the same standards of scrutiny before they too can become scientific fact.

Scientific mysteries are intriguing. Almost everyone loves a good mystery. Unlike fiction, these mysteries are real. Their reality is wondrous and sometimes scary. This book will "unravel" each mystery by presenting the currently held scientific theories to explain the observed phenomena. However, in the absence of a viable scientific explanation, when possible I will propose an explanation based on original research. Regardless of the origin of the explanations, please understand, we are on the edge of science where scientific proof is elusive, and scientific consensus is rare. Therefore, consider all such theories with an open, but cautious mind. Nobel Laureate Max Born said, "I am now convinced that theoretical physics is actually philosophy." Therefore, often the explanation will read like metaphysics or even science fiction. This is how life is on the edge of science, where mysteries abound.

I invite you to journey with me on the quest to unravel the mysteries of the universe, and to prepare for wherever that journey may lead us.

Section I

What Caused the Big Bang?

"A likely impossibility is always preferable to an unconvincing possibility."

—Aristotle (384-322 BC)

"Particles can 'pop up' out of a vacuum so long as they do not have too large a mass or do not last too long."

U.S. Department of Energy: Newton: Ask a Scientist "Quantum Fluctuations," 2004.

CHAPTER I

Something from Nothing

How did the universe begin? Did it even have a beginning, or is it eternal? Scientists and philosophers have been asking these questions for thousands of years. Theologians have been providing supernatural explanations that require a supreme being and, in several religions, numerous supreme beings. For example, Christians believe in one God, and in accordance with their belief, their God created the universe. The Egyptians, on the other hand, believed in many gods, and attributed the creation of the universe to them. However, in the early part of the Twentieth Century, a scientific answer began to emerge.

The entire question of the "birth" of the universe was brought into scientific focus when, in 1929, Edwin Hubble determined that the universe was expanding. The expanding-universe discovery led to what most scientists ascribe to as the Big Bang theory of the universe.

The Big Bang theory holds that the universe started 13.7 billion years ago as an infinitely dense energy point that expanded suddenly to create the universe. This is an excellent example of why the Big Bang theory belongs to the class of theories referred to as "cosmogonies" (theories that suggest the universe had a beginning). The Big Bang is widely documented in numerous scientific works, and is widely held as scientific fact by the majority of the scientific community.

The Big Bang theory provides an excellent framework of how the universe evolved, but it does not give us insight into what predated the Big Bang itself, or what caused it suddenly to go "bang." Indeed, these are two serious issues of the Big Bang theory, which are widely acknowledged by the scientific community.

Although the Big Bang has won the hearts and minds of most of the scientific community, other theories compete with the Big Bang. Of all the new theories, none has captured more attention than the multiverse theory. The multiverse theory is speculative, which means that it lacks direct experimental confirmation.

The multiverse theory holds that this universe is but one of a set of disconnected universes. There are numerous theories about the multiverse itself, which we will discuss in later chapters. None of the theories under serious consideration by the scientific community explains the origin of energy to create a Big Bang or a multiverse. The crucial question is deceptively simple. Where did the initial energy come from to fuel a Big Bang or create a multiverse? This is the largest mystery in science.

To unravel this mystery, we will start with an unusual phenomenon observed in the laboratory, namely spontaneous particle production or "virtual particles." The explanations below may become intimidatingly technical at times. Please do not be put off by the technical terms. Providing the scientific basis for virtual particles is crucial to understanding the next chapter. As you read on, most of your questions regarding the technical terms and the science will likely be resolved. You may consult the Glossary at the end of this book for further information on the technical terms and theories used throughout. You are not alone if you become confused. We are on the edge of science, where even scientists argue over the interpretation of observations and theories. With this in mind, we will continue with understanding spontaneous particle creation.

Spontaneous particle creation is the phenomenon of particles appearing from apparently nothing, hence their name "virtual particles." However, they appear real, and cause real changes to their environment. What is a virtual particle? It is a particle that only exists for a limited time. The virtual particle obeys some of the laws of real particles, but it violates other laws. What laws do virtual particles obey? They obey two of the most critical laws of physics, the

Heisenberg uncertainty principle (it is not possible to know both the position and velocity of a particle simultaneously), and the conservation energy (energy cannot be created or destroyed). What laws do they violate? Their kinetic energy, which is the energy related to their motion, may be negative. A real particle's kinetic energy is always positive. Do virtual particles come from nothing? Apparently, but to a physicist, empty space is not *nothing*. Said more positively, physicists consider empty space *something*.

Before we proceed, it is essential to understand a little more about the physical laws mentioned in the above paragraph.

First, we will discuss the Heisenberg uncertainty principle. Most physics professors teach it in the context of attempting to simultaneously measure a particle's velocity and position. It goes something like this:

- When we attempt to measure a particle's velocity, the measurement interferes with the particle's position.
- If we attempt to measure the particle's position, the measurement interferes with the particles velocity.
- Thus, we can be certain of either the particle's velocity or the particle's position, but not both simultaneously.

This makes sense to most people. However, it is an over simplification. The Heisenberg uncertainty principle has greater implications. It embodies the statistical nature of quantum mechanics. Quantum mechanics is a set of laws and principles that describes the behavior and energy of atoms and subatomic particles. This is often termed the "micro level" or "quantum level." Therefore, you can conclude that the Heisenberg uncertainty principle embodies the statistical behavior of matter and energy at the quantum level. In our everyday world, which science terms the macro level, it is possible to know both the velocity and position of larger objects. We generally do not talk in terms of probabilities. For example, we can predict the exact location and orbital velocity of a planet. Unfortunately, we are not able to make similar predictions about an electron as it obits the nucleus of an atom. We can only talk in probabilities regarding the electron's position and energy. Thus, most scientists will say that macro-level phenomena are deterministic, which means that a unique solution describes their state of being, including position, velocity, size, and other physical attributes. On the other hand, most physics will argue that micro level (quantum level) phenomena are probabilistic, which means that their state of being is described via probabilities, and we cannot simultaneously determine, for example, the position and velocity of a subatomic particle.

The second fundamental law to understand is the conservation of energy law that states we cannot create or destroy energy. However, we can transform energy. For example, when we light a match, the mass and chemicals in the match transform into heat. The total energy of the match still exists, but it now exists as heat.

Lastly, the kinetic energy of an object is a measure of its energy due to its motion. For example, when a baseball traveling at high velocity hits a thin glass window, it is likely to break the glass. This is due to the kinetic energy of the baseball. When the window starts to absorb the ball's kinetic energy, the glass breaks. Obviously, the thin glass is unable to absorb all of the ball's kinetic energy, and the ball continues its flight after breaking the glass. However, the ball will be going slower, since it has used some of its kinetic energy to break the glass.

With the above understandings, we can again address the question: where do these virtual particles come from? The answer we discussed above makes no sense. It is counter intuitive. However, to the best of science's knowledge, virtual particles come from empty space. How can this be true?

According to Paul Dirac, a British physicist and Nobel Prize Laureate, who first postulated virtual particles, empty space (a vacuum) consists of a sea of virtual electron-positron pairs, known as the Dirac sea. This is not a historical footnote. Modern-day physicists, familiar with the Dirac-sea theory of virtual particles, claim there is no such thing as empty space. They argue it contains virtual particles.

This raises yet another question. What is a positron? A positron is the mirror image of an electron. It has the same mass as an electron, but the opposite charge. The electron is negatively charged, and the positron is positively charged. If we consider the electron matter, the positron is antimatter. For his theoretical work in this area, science recognizes Paul Dirac for discovering the "antiparticle." Positrons and antiparticles are all considered antimatter.

Virtual particle-antiparticle pairs pop into existence in empty space for brief periods, in agreement with the

Heisenberg uncertainty principle, which gives rise to quantum fluctuations. This may appear highly confusing. A few paragraphs back we said that the Heisenberg uncertainty principle embodies the statistical nature of energy at the quantum level, which implies that energy at the quantum level can vary. Another way to say this is to state the Heisenberg uncertainty principle gives rise to quantum fluctuations.

What is a quantum fluctuation? It is a theory in quantum mechanics that argues there are certain conditions where a point in space can experience a temporary change in energy. Again, this is in accordance with the statistical nature of energy implied by the Heisenberg uncertainty principle. This temporary change in energy gives rise to virtual particles. This may appear to violate the conservation of energy law, arguably the most revered law in physics. It appears that we are getting something from nothing. However, if the virtual particles appear as a matter-antimatter pair, the system remains energy neutral. Therefore, the net increase in the energy of the system is zero, which would argue that the conservation of energy law remains in force.

No consensus exists that virtual particles always appear as a matter-antimatter pair. However, this view is commonly held in quantum mechanics, and this creation state of virtual particles maintains the conservation of energy. Therefore, it is consistent with Occam's razor, which states that the simplest explanation is the most plausible one, until new data to the contrary becomes available. The lack of consensus about the exact nature of virtual particles arises because we cannot measure them directly. We detect their effects, and infer their existence. For example, they produce the Lamb shift, which is a small difference in energy between two energy levels of the hydrogen atom in a vacuum. They produce the Casimir-Polder force, which is an attraction between a pair of electrically neutral metal plates in a vacuum. These are two well-known effects caused by virtual particles. A laundry list of effects demonstrates that virtual particles are real.

The above discussions distill to three key points. First, in accordance with the Heisenberg uncertainty principle, virtual particles pop in and out of existence in a vacuum. Second, we cannot measure virtual particles directly. Third, modern science believes virtual particles are real because they cause measurable changes to their environment.

This creation of virtual particles is sometimes termed spontaneous particle creation. Spontaneous particle creation raises an intriguing question. Are there hidden dimensions? Assume the Dirac sea model is correct, and that empty space (a vacuum) consists of a sea of virtual electron-positron pairs. If you are willing to accept this assumption, where are they located? It is a reasonable question. We are dealing with a vacuum, and at the same time asserting it contains electron-positron pairs. Where are they located? A possible explanation is they are in another dimension. As mind bending as this sounds, a formidable scientific theory known as M-theory asserts reality consists of eleven dimensions, not simply the four (three spatial, one temporal) we typically encounter. M-theory is "string" theory on steroids. At this point, I suspect you may be ready to blow a time-out whistle. This theory explains one puzzle using another puzzle.

Therefore, in the interest of clarity, we will take it one step at a time, and start by explaining more about M-theory. This will be a conceptual modeling of the theory.

In a sense, science has been working its way to M-theory since the discovery of atoms and subatomic particles, culminating in the discovery of the quarks (circa 1970s) as the fundamental building blocks for protons and neutrons. (Protons, neutrons, and electrons are the fundamental building blocks of atoms. Quarks are the fundamental building blocks of protons and neutrons.) In the 1980s, scientists claimed that these fundamental building blocks could be further reduced to infinitely small building blocks of vibrating energy, having only the dimension of length, termed "stings."

Conceptually, the "strings" vibrate in multiple dimensions. The vibration of the string determines whether it appears as matter or energy. According to string theory, every form of matter or energy is the result of the string's vibration.

By the 1990s, science recognized five different string theories, each with their own set of equations. The five string theories appeared valid, but scientists became uneasy. Surely, they could not all be right. In 1994, string theorist Edward Witten (Institute for Advanced Study), and other researchers, proposed a unifying theory called "M-theory." The "M" stands for "membrane." M-theory asserted that strings are one-dimensional slices of a two-dimensional membrane vibrating in eleven-dimensional space.

I understand it is hard, if not impossible, to picture an eleven-dimensional space because we live in a four-dimensional world. My picture goes something like this. The membrane (referred to as a "brane") is like a shadow of a million spread-out toothpicks. A shadow has two dimensions, and is the brane in this analogy. Each toothpick represents a string, having only the dimension of length. In this example, we are considering the toothpicks to have no width. Next, I think about this shadow being able to float off the surface and move around the room in three-dimensional space. It continually changes position in time. That is to say at time $t_{1,}$ it is in one place, and at another time $t_{2,}$ it is in another place. In this mind-bending analogy, we have accounted for seven dimensions. A two-dimensional shadow made from one-dimensional toothpicks accounts for three dimensions. The shadow floating in three-dimensional space accounts for three additional dimensions. Now, picture the shadow floating to a specific place at a specific time. When it moves to another place, time will have passed. The shadow, changing positions in time, accounts for one additional dimension (a temporal coordinate). How do I picture the other four? I think of there being small, invisible holes in space. The shadow can slip into, move around in, and disappear from view in these holes. The holes would represent a hidden three-dimensional space accounting for another three dimensions. The shadow moving in the holes would again represent another temporal coordinate. This analogy, which may be difficult to understand, is how I picture eleven-dimensional space. We live in a four-dimensional world. It is difficult to imagine seven other hidden dimensions.

Scientists, too, have a problem with the eleven-dimensional

model of reality that M-theory provides. The mathematics of M-theory is elegant, but correlating the mathematics to reality has frustrated numerous scientists. However, M-theory did accomplish one main goal. It unified the previous five spring theories into one. It demonstrated that each of the five was a specific case of M-theory. Well-known scientists, like Michio Kaku, Stephen Hawking, and Leonard Mlodinow, became proponents of M-theory, applauding its mathematical elegance, and suggesting it may be a candidate for The Theory of Everything. (The Theory of Everything would be a comprehensive scientific theory that explains the physical behavior of all matter and energy.) The one thing missing to make this picture perfect is experimental evidence. To date, we have no experimental evidence for M-theory. This does not mean M-theory is wrong or should be dismissed. Scientists continue to work on it, and experimental proof may eventually emerge.

"Fascinating," as Mr. Spock would say on Star Trek, but where does that leave us? Why am I bringing up M-theory and hidden dimensions? The answer is that spontaneous particle creation may have a connection to the hidden dimensions of M-theory. The entire Dirac sea (a vacuum filled with particle-antiparticle pairs) may exist in the hidden dimensions predicted by M-theory. Of course, it is easy for me, a theoretical physicist, to make this assertion since we have no proof of M-theory. However, we do have experimental evidence that enables us to infer that virtual particles exist. If they do exist, where are they located? Even if they exist as pure packets of energy (quanta), where are they located? One suggestion is to look into the hidden dimensions predicted by M-theory.

Are there hidden dimensions or is this science fiction? The scientific answer is: we don't know. However, as Edward Witten (American theoretical physicist) said, "As far as extra dimensions are concerned, very tiny extra dimensions would not be perceived in everyday life, just as atoms are not: we see many atoms together but we do not see atoms individually." We know atoms exist, but we cannot see them. Could this be true of hidden dimensions? How do we experimentally prove the hidden dimensions of M-theory? Currently, scientists are using the largest particle colliders to create near speed-of-light collisions between subatomic particles. To understand this approach to prove hidden dimensions, we need to understand what is occurring when a particle with a mass is accelerated near the speed of light, resulting in a relativistic kinetic energy (energy due to its motion). The total mass-energy of the accelerated particle is equal to the mass plus the relativistic kinetic energy. By causing two particles of known mass-energy to collide, they are able to determine if the sum of all the mass-energy before the collision equals the mass-energy after the collision. Two important laws are utilized to make this calculation. Einstein's famous mass-energy equivalence ($E=mc^2$, where E is energy, m is mass, and c is the speed of light in a vacuum), and the conversation of energy law (which states energy cannot be created nor destroyed). By painstakingly accounting for all of the mass-energy before the collision to the mass-energy after the collision, they are able to look for missing mass-energy. If they find such a result, it could imply additional dimensions. That is to say, the mass-energy went into another dimension. These experiments continue as I write. The next few years should be very exciting.

This brings up a crucial question that may have already occurred to you. Could the Big Bang itself be the result of a quantum fluctuation, similar to how virtual particles form? We will scientifically examine that possibility in the next chapter.

"People need to be aware that there is a range of models that could explain the observations... What I want to bring into the open is the fact that we are using philosophical criteria in choosing our models. A lot of cosmology tries to hide that."

Astrophysicist George F. R. Ellis
"Profile: George F. R. Ellis,"
Scientific American, October 1995

CHAPTER 2

The Big Bang—Singularity or Duality?

The best-known proponent of the idea that a quantum fluctuation gave birth to the energy of the Big Bang is Canadian-American theoretical physicist, Lawrence Maxwell Krauss. Dr. Krauss is a Professor of Physics at Arizona State University, and the author of several bestselling books, including: *A Universe from Nothing: Why There is Something Rather Than Nothing* (Free Press, 2012). In the simplest terms, Dr. Krauss ascribes the creation of the universe to a quantum fluctuation, similar to how virtual particles gain existence. An excellent presentation of his theory is on YouTube, titled: "A Universe From Nothing" by Lawrence Krauss, AAI 2009. The hour-long YouTube video has over a million views, making it one of their most popular videos. Dr. Krauss is a gifted physicist with the ability to explain difficult concepts simply and entertainingly. Dr. Krauss' book and video essentially make the same points. His main hypothesis is that the Big Bang is the result of a quantum fluctuation, much like a virtual particle, except on a cosmic scale.

As stated earlier, a quantum fluctuation results when a point in space experiences a temporary change in energy. This behavior is due to the Heisenberg uncertainty principle, which delineates the statistical nature of matter and energy at the level of atoms and subatomic particles. For an analogy, consider heating a large building. Some parts of the building will be warmer than others. Statistically, the warm air does not disperse evenly throughout the building. This means the energy (warm air) in the building varies. This same phenomenon occurs at the atomic and subatomic level. The energy in a vacuum, predicted by the Dirac sea (which postulates a vacuum is filled with electron-positron pairs), statistically becomes unevenly distributed. This temporary change in energy gives rise to virtual particles.

I found Dr. Krauss' hypothesis convincing, especially in light of what we observe regarding virtual particles. However, one intriguing aspect about virtual particles is that we sometimes observe their occurrence in matter-antimatter pairs. This raised a question. Why would the Big Bang "particle" be a singularity? In this context, we can define a "singularity" as an infinitely energy-dense particle. Numerous observations about virtual particles suggest a "duality." A "duality," in this context, would refer to an infinitely dense energy particle pair (one matter particle, and the other an antimatter particle). How would all this play out?

First, we need to postulate a super-universe, one capable of quantum fluctuations. Cosmologists call the super-universe the "Bulk." The Bulk is "empty" space, which gives existence to infinitely energy-dense matter-antimatter virtual particles. These collide and initiate the Big Bang. If this view of reality is true, it makes the multiverse concept more plausible. Other infinitely energy-dense matter-antimatter particles continually pop in and out of existence in the Bulk, similar to the way that virtual matter-antimatter particles do in the laboratory. When this occurs in the Bulk, a collision between the particles initiates a Big Bang. Therefore, considering the billions of galaxies in the universe, there may be billions of universes in the Bulk.

Next, we'll summarize this theory, which I call The Big Bang Duality. It has two major hypotheses, which I will delineate below, along with rationale:

Hypothesis 1: A super-universe, the "Bulk," is capable of producing quantum fluctuations, in accordance with the Heisenberg uncertainty principle. The quantum fluctuations result in infinitely energy-dense particle-antiparticle pairs. *Rationale:* The quantum-fluctuation theory of the Big Bang is entirely plausible. Quantum mechanics and the virtual-particle theory provide a solid foundation for it. It is reasonable to consider that the quantum fluctuation in the Bulk resulted in an infinitely energy-dense particle-antiparticle pair, not a single energy-dense particle. Why is this reasonable? In laboratory observations, high-energy vacuum environments favor the production of virtual particle pairs. This equates to a neutral system, which aligns with the conservation of energy and momentum. The notion that the Bulk gives rise to infinitely energy-dense virtual particles suggests that the Bulk itself is highly energized. Therefore, the Bulk would be expected to give rise to infinitely energy-dense virtual particle-antiparticle pairs.

Hypothesis 2: The Big Bang results when the energy-dense particle-antiparticle pair collides in the Bulk. *Rationale:* This is essentially the ignition point of the Big Bang. The elegance of this theory is that it explains the Big Bang's initiation, and the absence of antimatter. Both elements are subjects of later chapters.

One can draw several remarkable inferences from the Big Bang Duality theory.

1) The entire concept that the universe is the result of a quantum fluctuation suggests there are "natural" laws of quantum mechanics that apply in the Bulk (the super-universe). This makes a strong case that the scientific laws of the universe originate in the Bulk. This implies the physical laws of the universe pre-date the Big Bang, and if there were other universes created via quantum fluctuations, they too would obey the laws of the Bulk.

2) Since we see virtual particle pairs continually pop in and out of existence in an energized laboratory vacuum, there is reason to believe a similar process occurs in the Bulk. This suggests other universes originate in the Bulk, and follow the laws of the Bulk. Therefore, other universes in the Bulk conceivably look like ours, and are evolving similarly. If we create computer models of the Big Bang and change a fundamental physical law, we end up with a failed universe. For example, if gravity is much greater or weaker than it is in our universe, the computer simulation of the universe from a Big Bang fails. We end up with a failed universe. This argues that if there are other universes, they most likely look like ours—or they would fail as a universe.

The Big Bang Duality theory appears to fit observations regarding virtual particles. It strongly suggests that the Big Bang was a duality, not a singularity. It also suggests that the physical laws of our universe originate in the Bulk, and pre-date the Big Bang. Einstein often remarked, "As I have said so many times, God doesn't play dice with the world." My reading of this is that Einstein believed in an underlying order to the universe. Einstein remarked, "Try and penetrate with our limited means the secrets of nature and you will find that, behind all the discernible concatenations, there remains something subtle, intangible, and inexplicable. Veneration for this force beyond anything that we can comprehend is my religion. To that extent I am, in point of fact, religious."

To my mind, Einstein is arguing that the universe has is an underlying order, and this aligns with the Big Bang Duality assertion that the scientific laws of our universe originate in the Bulk, and pre-date the Big Bang.

One last compelling point: the Big Bang Duality theory provides a basis to postulate other universes. This lends credence to the multiverse theories, discussed later. We know from laboratory experiments that virtual matter-antimatter pairs pop in and out of existence continually. This suggests it is reasonable to assume a similar process occurs in the Bulk. Virtual particles in the Bulk, in the form of infinitely energy-dense matter-antimatter pairs, are continually popping in and out of existence. When they collide, they initiate a Big Bang, and form new universes. Later we will return to visit the Big Bang Duality theory in order to explain why the universe consists of matter, not antimatter.

The majority of the scientific community accepts the Big Bang theory, namely the universe originated from an infinitely dense-energy point. Above we discussed potential theories to explain how the infinitely dense energy point came to exist. However, a serious question remains unanswered. Why did it go bang? This profound mystery is the subject of our next chapter.

"Energy in any form seeks stability at the lowest energy state possible, and will not transition to a new state unless acted on by another energy source."

Louis A. Del Monte, Physicist
Minimum Energy Principle
Unraveling the Universe's Mysteries (2012)

CHAPTER 3

What Made the Big Bang Go Bang?

This is a little play on words. The Big Bang theory holds that the evolution of the universe started with an infinitesimal packet of near infinite energy (termed a "singularity") that suddenly expanded and continues to expand. If this is true, was it big? No, it was infinitesimally small. Did it go bang? No, it expanded. Space is a vacuum, and it is unable to transmit sound waves. Therefore, there were no sound waves to make a bang noise. Granted, I was not there since it took place 13.7 billion years ago, and you are certainly entitled to your own opinion. I am only jesting, but the description above of the Big Bang theory is what the scientific community holds to be responsible for the evolution of the universe.

In the last two chapters, we discussed the origin of the energy that led to the Big Bang. However, we did not discuss one big question. What initiated the Big Bang's expansion?

Throughout the theories of science, there appears to be a common thread based on well-observed physical phenomena regarding the behavior of energy. That common thread states that differences in temperature, pressure, and chemical potential always seek equilibrium if they are in an isolated physical system. For example, with time, a hot cup of coffee will cool to room temperature. This means it reaches equilibrium (balance, stability and sameness) with the temperature of the room, which is the isolated physical system in this example. Readers familiar with thermodynamics will instantly attribute this behavior of energy as following the second law of thermodynamics. However, the same law, worded differently, exists in numerous scientific contexts. In the interest of clarity, I am going to restate this law, describing the behavior of energy, in a way that makes it independent of scientific contexts. In a sense, it abstracts the essence of the contextual statements, and views applications of the law in various scientific contexts as specific cases. I am not the first physicist to undertake generalizing the second law of thermodynamics to make it independent of scientific contexts. However, I believe my proposed restatement provides a simple and comprehensive description of the laws that energy follows, and it will aide in understanding concepts presented in later chapters. For the sake of reference, I have termed my restatement the Minimum Energy Principle.

Energy in any form seeks stability at the lowest energy state possible, and will not transition to a new state unless acted on by another energy source.

Consider these two examples to illustrate the Minimum Energy Principle.

1) Radioactive substances. Radioactive substances emit radiation until they are no longer radioactive (they become stable). However, by introducing other radioactive substances under the right conditions, they can transition to a new state. Indeed, if the proper radioactive elements combine under the right circumstances, the result can be an atomic explosion.

2) A thermodynamic example. Consider a branding iron fresh from the fire. It emits thermal radiation until it reaches equilibrium with its surroundings. In other words, once a branding iron leaves the fire, it will start to cool by transferring energy to its surrounding. Eventually, it will be at the same temperature as its surroundings. (This illustrates the first part of the Minimum Energy Principle: Energy in any form seeks stability at the lowest energy state possible.) However, if we increase the temperature of the branding iron by placing it back in the fire, the branding iron will absorb energy until it again reaches equilibrium with the temperature of the fire. (This illustrates the second

part of the Minimum Energy Principle: It transitions to this new state by being acted on by the fire. The fire acts as an energy source.)

The Minimum Energy Principle is consistent with the law of entropy. To understand this, we will need to discuss entropy. In classical thermodynamics, entropy is the energy unavailable for work in a thermodynamic process. For example, no machine is one hundred percent efficient in converting energy to work. A portion of the energy is always lost in the form of waste heat. An example is the miles per gallon achievable via your car engine, ignoring other factors such as the weight of the vehicle, its aerodynamic design, and other similar factors. Several car manufacturers are able to build highly efficient engines. However, no car manufacturer can build an engine that is one hundred percent efficient. As a result, a fraction of total energy is always lost, typically in the form of waste heat.

Entropy proceeds in one direction, and is a measure of the system's disorder. Any increase in entropy implies an increase in disorder and an increase in stability. For example, the heat lost in a car engine is lost to the atmosphere, and is no longer usable to do work. The heat lost is adding to the disorder of the universe, and is a measure of entropy. Oddly, though, the lost heat is completely stable.

In a given system, entropy is either constant or increasing, depending on the flow of energy. If the system is isolated, and has no energy flow, the entropy remains constant. If the system is undergoing an energy change, such as ice melting in a glass of water, the entropy is increasing. When the ice completely melts, and the system reaches equilibrium with its surrounding, it is stable. This has a significant implication. Entropy is constantly increasing in the universe since everything in the universe is undergoing energy change. In theory, the entropy of the universe will eventually maximize, and all reality will be lost to heat. The universe will be completely stable and static. I have termed this the "entropy apocalypse." I know I am being a little dramatic here, but most of the scientific community believes the entropy (disorder) of the universe is increasing. Eventually, all energy in the universe will be stable and unusable, all change will cease to occur, and the universe will have reached the entropy apocalypse.

Based on the above discussion of entropy, we can argue that entropy seeks to maximize and, therefore, reduce energy to the lowest state possible. This is why I stated that the Minimum Energy Principle, which asserts that energy seeks the lowest state possible, is consistent with law of entropy.

How does this help us understand what made the Big Bang go bang? The Minimum Energy Principle, along with our understanding of the behavior of entropy, makes answering this question relatively easy. The scientific community agrees that the Big Bang started with a point of infinite energy, at the instant prior to the expansion. From the Minimum Energy Principle, we know "Energy in any form seeks stability at the lowest energy state possible and will not transition to a new state unless acted on by another energy source." Since we know it went "bang," we can make three deductions regarding the infinitely dense-energy point. First, it was not stable. Second, it was not in the lowest energy state possible. Third, the entropy of the infinitely dense-energy point was at its lowest state possible, which science terms the "ground-state entropy." These three conditions set the stage for the Minimum Energy Principle and the laws of entropy to initiate the Big Bang.

By the very nature of "playing the tape" of the expanding universe back to discover its origin, namely the Big Bang, we can conclude a highly dense energy state. It will be a highly dense energy state because we are going to take all the energy that expanded from the Big Bang, and cause it to contract. As it contracts, the universe grows smaller and more energy-dense. At the end of this process, we have a highly dense energy state. I think of it as a point, potentially without dimensions, but with near-infinite energy. This view is widely held by the scientific community. If it is true, all logic causes us to conclude it was an infinitely excited energy state, and we would have every reason to question its stability—and to believe it was at the "ground-state" entropy (the lowest entropy state possible).

Our observations of unstable energy systems in the laboratory suggest that as soon as the point of infinite energy came to exist, it had to seek stability at a lower energy level. The Big Bang was a form of energy dilution. In the process of lowering the energy, it increased the entropy of the universe. Once again, we see the Minimum Energy Principle and the law of entropy acting in concert.

How long did the infinitely dense-energy point exist? No one really knows. However, we can approach an answer by understanding more about time.

Discussing the Big Bang in terms of time, as we typically understand time, is difficult. It will not do any good to look at your watch or think in small fractions of a second. Stop-motion photography will not work this time. Those times are infinitely large compared to Planck time (~ 10^{-43} seconds, which is a one divided by a one with forty-three zero after it). Theoretically, Planck time is the smallest timeframe we will ever be able to measure. So far, we have not even come close to measuring Planck time. The best measurement of time to date is of the order 10^{-18} seconds.

What is so significant about Planck time? The fundamental constants of the universe formulate Planck time, not arbitrary units. According to the laws of physics, we would be unable to measure "change" if the time interval were shorter that Planck time. In other words, Planck time is the shortest interval we humans are able to measure, or even comprehend change to occur. Scientifically, it can be argued that no time interval is shorter that Planck time. Thus, the most rapid change can only occur in concert with Planck time, and no faster. Therefore, when we discuss the initiation of the Big Bang, the smallest time interval we can consider is Planck time.

The whole notion of Planck time, and its relationship to the Big Bang, begs another question. Did time always exist? Most physicists say NO. Time requires energy changes, and that did not occur until the instant of the Big Bang. Stephen

Hawking, one of the world's most prominent physicists and cosmologists, is on record that he believes time started with the Big Bang. Dr. Hawking asserts that if there was a time before the Big Bang, we have no way to access the information. However, an argument can be made that time pre-dates the Big Bang. How is this possible?

If we consider the Big Bang is the result of a quantum fluctuation in the Bulk, energy changes are occurring in the Bulk. This implies time exists in the Bulk and pre-dates the Big Bang. This begs the question: is there any evidence of a Bulk and other universes? A growing number of scientists say YES. They cite evidence that our universe bumped into other universes in the distant past. What is the evidence? They cite unusual ring patterns on the cosmic microwave background. The cosmic microwave background is leftover radiation from the Big Bang, and is the most-distant thing we can see in the universe. It is remarkably uniform, with the exception of the unusual ring patterns. Scientists attribute the ring patterns to bumps from other universes. Two articles discuss this finding.

- *First evidence of other universes that exist alongside our own after scientists spot "cosmic bruises,"* by Niall Firth, December 17, 2010 (http://www.dailymail.co.uk).
- *Is Our Universe Inside a Bubble? First Observational Test of the "Multiverse."* ScienceDaily.com, August 3, 2011.

Obviously, this is controversial, and even the scientist involved caution the results are initial findings, not proof. It is still intriguing, and lends fuel to the concept of there being other universes. This would suggest time, in the cosmic sense, transcends the Big Bang. As impossible as it would seem to prove other universes, science has found ways of proving similar scientific mysteries. The prominent physicist, Michio Kaku, said it best in *Voices of Truth* (Nina L. Diamond, 2000), "A hundred years ago, Auguste Compte, ... a great philosopher, said that humans will never be able to visit the stars, that we will never know what stars are made out of, that that's the one thing that science will never ever understand, because they're so far away. And then, just a few years later, scientists took starlight, ran it through a prism, looked at the rainbow coming from the starlight, and said: 'Hydrogen!' Just a few years after this very rational, very reasonable, very scientific prediction was made, that we'll never know what stars are made of." This argues that what seems impossible to prove today might be a scientific fact tomorrow.

A theoretical case argues that cosmic time in the Bulk predated the Big Bang. Eventually we may be able to prove it. It is reasonable to believe time for our universe started with the Big Bang. This is our reality. This is consistent with Occam's razor, which states the simplest explanation is the most plausible one (until new data to the contrary is available). For our universe, the Big Bang started the clock ticking, with the smallest tick being Planck time.

We are finally in a position to answer the two crucial questions. First, what made the big bang go bang? Second, how long did the infinitely dense energy point exist before it went bang?

Why did the Big Bang go bang?

The Big Bang followed the Minimum Energy Principle, "Energy in any form seeks stability at the lowest energy state possible, and will not transition to a new state unless acted on by another energy source." The infinitely dense energy point, which science terms a "singularity," sought stability at the lowest energy state possible. If it was "duality," as argued in Chapter 2, the collision of the infinitely energy-dense matter and antimatter particles would represent the unstable infinitely energy-dense state. Therefore, the arguments presented apply equally to a "singularity" or "duality." Being infinitely energy-dense, implies instability and minimum entropy (ground-state entropy). Thus, it required dilution to become stable, which caused entropy to increase. The dilution came in the form of the "Big Bang." Since we were dealing with an unstable infinitely energy-dense point, the Big Bang went bang at the instant of existence. The instant of existence would correlate to the smallest time interval science can conceive, the Planck time. This process is continuing today as space continues its accelerated expansion.

This gives us a reasonable explanation of why the Big Bang went bang. It argues that it went "bang" at the exact instant it came to exist. However, the Big Bang is not the sole theory regarding the evolution of the universe. As I mentioned in previous chapters, it has a competitor, namely the multiverse theories. Unlike the Big Bang, though, the multiverse theories lack a solid scientific foundation. However, they have captured the imaginations of millions throughout the world, including scientists and nonscientists alike. A growing number of scientists assert the multiverse theories offer a credible alternative to the Big Bang. In addition, elements of the multiverse theories address the crucial question: What happened before the Big Bang? I think you will find discussing multiverse theories fascinating in the next chapter.

"It's quite striking to me that the mathematically simplest theories tend to give us multiverses. It's proven remarkably hard to write down a theory which produces exactly the universe we see and nothing more."

Scientific American, July 19, 2011,
The Case for Parallel Universes, Max Tegmark

CHAPTER 4

The Multiverse Theories

You may think we are about to discuss something brand new. However, American philosopher and psychologist William James actually coined the term "multiverse" in 1895. If only he knew that over a hundred years later scientists, philosophers, and science fiction writers would still be using the term.

Ironically, the term multiverse is self-contradictory! The term universe is from the Latin, and means "entirety." How is it possible to have more than the entirety? We scientists have adopted the term, and have used it to describe all the possible universes that could exist, including our own. Sometimes, instead of using the term "multiverse," the phrases "parallel universes," "alternative universes," "quantum universes," "parallel dimensions," and "parallel worlds," find application in physics, cosmology, philosophy, and science fiction, to name a few of the contexts.

We have four main multiverse hypotheses (four hypothetical possible universes, including the universe we currently experience, that together comprise everything that exists). We will discuss each in detail. You will discover the nature of each universe, within the multiverse, and their relationship to each other.

Let me be clear about one salient point. Everything we are about to discuss regarding the multiverse is hypothetical, with scant observable evidence. We have no conclusive proof of the existence of any universe beyond the one we currently experience. Nonetheless, several prominent cosmologists, such as Max Tegmark, believe it is inferable from other prominent theories of science, including cosmic inflation, quantum mechanics, and string theory—all of which are provable. Often, scientists use multiverse theory to address the unexplained questions that the Big Bang theory, viewed in isolation, does not answer. Although the hypotheses of a multiverse may contradict current paradigms of the universe, it deserves consideration. When Einstein proposed his special theory of relativity in 1905, it too lacked proof, and thus skepticism abounded. By 1920, the special theory of relativity had become widely accepted by the scientific community. From 1905 through today, numerous experiments have validated the fundamental elements of the special theory of relativity. In my opinion, advances in science require healthy skepticism, debate and proof. We are at the point of healthy skepticism and debate regarding the multiverse. For the multiverse to be elevated to science fact, experimental proof will be required. The new super colliders may give us that proof in the next particle collision.

To assure that the four main theories of the multiverse receive appropriate consideration, each will receive its own section, showing them in the order developed by cosmologist Max Tegmark. The order he developed (Level I-IV) is a brick-by-brick approach. It enables each of the subsequent levels to build and expand on the previous levels.

Level I Multiverse: Beyond Our Cosmological Horizon

What is a cosmological horizon? This may be a hard concept to understand, so we'll walk our way through it. The universe is about 13.7 billion years old. However, the universe is larger than 13.7 billion light years in diameter due to the expansion and subsequent inflation of space, explained in the discussion of the Big Bang theory. In fact, our best current estimate, taking expansion and inflation of space into account, puts the edge of the observable universe at about 46–47 billion light-years away from Earth. This "edge" would represent our current cosmological horizon.

The Level I multiverse theory essentially assumes an infinite universe. The edge of the Level I multiverse is not

46-47 billion light-years from Earth, but extends an infinite distance from Earth. Scientists have termed this infinite universe a "super-universe." If the infinite universe theory is correct, our universe may be one universe out of uncountable billions in the super-universe. We cannot to see the other universes because our current observation technology is unable to look through the cosmic microwave-background radiation, which originated when the matter in the universe was plasma (hot, ionized gas), and thus opaque. In theory, if we develop more advanced observation technology, such as a neutrino telescope (one capable of detecting neutrinos) or even a gravitational telescope (one capable of detecting the yet-undiscovered gravitation particle called a "graviton"), we would be able to look beyond the cosmic microwave-background radiation and see older events. We would have a new cosmological horizon, but we would never be able to examine the "edge" of an infinite universe. Why? It has no edge—and advances in cosmic observation technology will not matter. Even the hypothetical graviton (the theoretical particle of gravity), traveling at the speed of light, would never reach us from an infinitely distant universe.

Why is an infinite universe even plausible? We know from actual observations that the universe's expansion is accelerating. The farther out our instruments allow us to observe, we can measure that the expansion is accelerating, and even exceeding the speed of light. The accelerating expansion is termed "inflation," and was confirmed in the late 1990s, as explained earlier when discussing the Big Bang theory. Until inflation's confirmation, scientists believed that gravity would eventually slow the universe's expansion, and even eventually cause the universe to contract in a "Big Crunch," since gravity causes everything to pull on everything.

Long before we had any observable proof of the universe's inflationary expansion, two scientists independently postulated its existence in 1979. Unfortunately, one scientist, Alexei Starobinsky of the L.D. Landau Institute of Theoretical Physics in Moscow, was unable to communicate his work to the worldwide scientific community due to the political policies of the former Soviet Union. Fortunately, the other scientist, Alan Guth, Professor of Physics at the Massachusetts Institute of Technology, developed an inflationary model independently, and communicated it worldwide. Guth's model, however, was not able to reconcile itself with the isotropic, homogeneous universe we observe today. In other words, to the best of our current ability to measure it, the universe essentially looks the same in every direction. Andrei Linde, Russian-American theoretical physicist and Professor of Physics at Stanford University, solved Guth's theoretical dilemma in 1986. Linde published an alternative model entitled "Eternally Existing Self-Reproducing Chaotic Inflationary Universe" (known as "Chaotic Inflation theory"). In Linde's model, our universe is one of countless others. A prediction of the chaotic inflation theory is an infinite universe with bubble universes within it. Would they be the same as our universe? No one knows. Perhaps one or more universes would be different from ours. However, being infinite, an infinite number of universes would be identical to ours, even down to the last atom, obeying the same physical laws.

The Level I multiverse is a mind-bending theory, but it has reasonable theoretical science at its core. If we extend the Level I multiverse to the limit, an infinite number of us (you, me, and everyone else) are out there somewhere beyond the cosmic horizon. Given an infinite number of us, we are living out every possible scenario. This is difficult to comprehend because infinite numbers cannot be comprehended. Here is a simple way to think about this. If you play poker, what are the odds that you will be dealt a royal flush (Ace, King, Queen, Jack, Ten, all in the same suit) in the first five cards? They are 2,598,960 to 1. That means you will get a royal flush about once every 2,598,960 hands of five-card poker (known as five-card stud poker). Even if you play every day, and for numerous hours a day, you may never get one. However, if you have forever, and continue playing, eventually you will get one, then another, and with infinite time, an uncountable number (an infinite number). Using this example, if there are an infinite number of us in the universe, then each of us in some part of the universe will experience a possible scenario. Since there are an infinite number of us, as a group we will experience every conceivable scenario. For example, in one of these possible scenarios, you would be the President of the United States.

Skeptical? I think skepticism is healthy. I encourage you to be skeptical until we have concrete evidence of an infinite universe. With that in mind, we will move on to the next level in the multiverse hierarchy.

Level II Multiverse: Bubble Universes

In our earlier discussion of Chaotic Inflation theory, we stated that a prediction of chaotic inflation theory is an infinite universe with bubble universes within it. Why is this?

Chaotic Inflation theory asserts the multiverse is stretching, and will continue stretching forever, but not uniformly. Portions of the multiverse stop stretching, while other portions continue to stretch. (In this context, the word "stretching" and "inflation" are equivalent.) This causes the stretched portion to break away, much like a small soap bubble can break away from a larger one. The portions of the multiverse that break away are embryonic Level I universes. Here is an easy way to think about it. When you make bread, you add yeast to the dough. The entire dough begins to expand. Inside the dough are small bubbles. In this analogy, the multiverse is the ever-expanding loaf of bread. The embryonic Level I universes are the bubbles inside the loaf of bread. How many bubble universes are there? No one knows for sure. Linde and Vanchurin calculated the possible number of bubble universes (source: Zyga, Lisa "Physicists Calculate Number of Parallel Universes," PhysOrg, 16 October 2009). Their calculation resulted in an incredibly large number of bubble universes (ten billion raised to the ten million power). If you try writing that number out without using exponentials ($10^{10^{10,000,000}}$, 10 raised to the power of ten raised to the power of ten million), you would be at it for a lifetime. Why is it incredibly big? We started with the assumption of an infinite universe capable of making an infinite number of bubble Level I universes. Anytime you deal with infinite numbers, it is hard to calculate something finite from something infinite. How accurate is this calculation? I leave it to you to decide. Suffices to say, I have my doubts, but will keep an open mind. Are all the bubble universes the same? Again, no one knows. If you assume a hypothetical unproven process termed "spontaneous symmetry breaking," which enables each new embryonic universe to be different from every other embryonic universe, one or even all could be different. If they are different, different physical laws may apply. The physical constants we hold sacred, like the speed of light in a vacuum, could conceivably be different. The Level II bubble universes are not scientifically well grounded. They require three assumptions that have no scientific verification:

1) An infinite universe expands, but not uniformly
2) Portions of that universe break away to form a new embryonic universes (bubble universes)
3) A hypothetical unproven process, termed spontaneous symmetry breaking, enables each new embryonic universe to be different from every other embryonic universe. Thus, each can have its own physical laws and constants.

Obviously, there are problems with the bubble universes theory. It stretches the limits of credibility. So, let's move on to the next level multiverse, which has significant support in the scientific community.

Level III Multiverse: The Many-Worlds Interpretation of Quantum Mechanics

We are about to discuss the unbelievable "many-worlds interpretation of quantum mechanics." Yet, according to a poll by "political scientist" L. David Raub, involving 72 "leading cosmologists and other quantum field theorists" (circa-1988), it is true. According to another relatively recent poll published in "The Physics of Immortality" (1994), 58% of scientists believe the many-world interpretation of quantum mechanics is true, 13% are on the fence (undecided), 11% have no opinion, and 18% do not believe it.

The scientists who appear to be among the believers are those described as string/brane theorists or quantum gravitists/cosmologists. They include Stephen Hawking, and Nobel Laureates Murray Gell-Mann and Richard Feynman. The majority of mainstream scientists, specializing in other areas of science, are mostly ignorant of it or do not have an opinion. However, the majority of those closest to the science behind the many-worlds theory are believers. I thought, prior to explaining a theory that will sound unbelievable, it best to provide insight into why you should even bother to read this chapter.

We will start with a simplified understanding of quantum mechanics. The simplest way of thinking about quantum mechanics is to compare it to Newtonian mechanics, which consists of Newton's laws of motion. On a macro-level (our everyday world), Newton's laws provide an excellent way to describe and predict the behavior of physical objects. However, on an atomic level, Newton's laws are unable to explain the behavior of physical objects. This led to the development of the new science of quantum mechanics. For clarity, we can define quantum mechanics as the branch of physics that mathematically describes the motion and interaction of atomic and subatomic particles.

To understand the many-world interpretation of quantum mechanics, we will take it one-step at a time.

First, numerous equations and mathematical models, developed primarily in the first half of the Twentieth Century, are successfully utilized to explain and predict the behavior of physical objects at the atomic and sub-atomic level. The equations explain and predict observed phenomena. However, the equations are complicated, and often require counterintuitive assumptions. For example, one fundamental assumption is that all physical objects have a wave-particle duality, describable mathematically via a "wavefunction." In the context of quantum mechanics, the wavefunction thoroughly describes the physical system. Since we do not directly experience the atomic level, interpreting the models and equations has been difficult. Scientists have asked whether the wavefunction is real—or is it a mathematical construct. Is quantum mechanics deterministic (the same physical conditions will always result in the same physical effect)? How do the mathematics and reality relate? This is where the problems begin to arise, bringing a number of contending schools of thought. We will discuss the two most significant schools of thought:

1) The Copenhagen interpretation: This is what physic professors typically teach in college-level quantum

mechanics classes. Niels Bohr, Werner Heisenberg and others, collaborating over a three-year period (1924–27), first proposed it. The Copenhagen interpretation holds that quantum mechanics does not provide a description of a single-objective reality. That means it is not deterministic. The equations, and mathematical models, result in probabilities of measuring various aspects of energy. Further, the act of making a measurement causes an unusual effect. The set of probabilities immediately and randomly assume one of the possible values. Quantum physicists describe this phenomenon as "wavefunction collapse." For clarity, we will illustrate the Copenhagen interpretation with an example: When we apply quantum equations to calculate the energy level of an electron in an atom, we get results bounded by probabilities, and not a specific energy level. If we measure the electron's energy level, we get a specific value (not probabilities). The Copenhagen interpretation asserts that the wavefunction describing the electron's energy level collapsed to a specific value.

2) Hugh Everett's many-worlds interpretation: Similar to the Copenhagen interpretation, it holds that certain observations are not predictable absolutely. This means, in agreement with the Copenhagen interpretation, there is a range of possible observations/results associated with physical phenomena, each associated with a different probability. However, Everett's interpretation is that

> each possible observation corresponds to a different universe, hence the name "many-worlds."

For clarity, we will illustrate Everett's many-worlds interpretation with an example. Consider what happens when you toss a coin up in the air. Two outcomes are possible. When it lands on the ground, it could land on heads— or tails. The Everett many-worlds interpretation contends it did both simultaneously, but each happened in a different universe. If you observe "heads," you are in one universe. Another you observe "tails," but that you is in a different universe. Neither of you know about the other, nor can you influence the other.

The many-worlds interpretation reads and sounds like science fiction. Yet, as I stated at the outset of this chapter, a number leading cosmologists and other quantum field theorists believe it is science fact, not science fiction. This still leaves us with an enormous unanswered question. Where does the energy come from to generate these worlds? We could argue that the origin of the energy is similar to the origin of the energy of the Big Bang. In earlier chapters, we discussed that it originated via the collisions of matter-antimatter infinitely energy-dense virtual particles in the Bulk. The basic problem is the Big Bang is widely accepted as science fact, because it grounded in scientific measurements of our universe. The many-worlds theory is simply a theory. We have no scientific evidence of other worlds. We are unable to observe other worlds. Yet, the many-worlds theory is hard to ignore when several of the greatest minds in science believe it is true.

If you are still with me in this universe, we are ready to discuss the last multiverse level.

Level IV Multiverse: The Ultimate Ensemble

Why is this the "last" multiverse in the Tegmark hierarchy? Because it is Dr. Tegmark's hierarchy, and he determined the order. I am being a bit flippant. However, what I have said has a grain of truth. The Ultimate Ensemble is Dr. Tegmark's hypothesis. The nature of the hypothesis precludes there being a Level V multiverse.

Dr. Tegmark's central hypothesis, for the Ultimate Ensemble, is that all structures that exist mathematically exist physically. That's it. One hypothesis. However, the implications are vast. If it can be mathematically modeled, it should be considered real. Therefore, if you solve an equation for a physical reality, such as the energy level of an electron in an atom, and it has two results that satisfy the equation, it implies two realities. In fact, all universes described by any mathematical structure are real, according to this theory. This opens the door to worlds corresponding to different sets of initial conditions, physical constants, and altogether different equations.

Probably sounds far out, and you may think this is New-Age thinking. However, it is a form of Platonism (the philosophy of Plato, dating back to ~ 400 BC), in that it postulates the physical existence of mathematical entities.

Why is this the last multiverse theory? According to Dr. Tegmark, if it is possible to express it as a mathematical model, it exists. This, according to Dr. Tegmark, "implies that any conceivable parallel universe theory can be described at Level IV," hence the name Ultimate Ensemble.

So, are we done with the multiverse? Did Dr. Tegmark's assertion that Level IV is the end of multiverse theories hold up? What do you think? You probably already know that the minds of scientists refuse boundaries. The reality is that not everyone agrees, which brings us to yet another type of multiverse. Some scientists consider this next multiverse, known as M-theory, to fit inside the Level IV multiverse, as described by Dr. Tegmark. Some do not. What is an M-theory universe? Read on.

"According to M-theory, ours is not the only universe. Instead, M-theory predicts that a great many universes were created out of nothing." [and] "...these multiple universes arise naturally from physical law."

Stephen Hawking, The Grand Design, 2010

CHAPTER 5

M-theory (Tying the String Theories Together)

M-theory was discussed briefly in Chapter 1 in the context of the Dirac sea model, which holds that empty space consists of a sea of virtual electron-positron pairs. We asked a significant question: where are they located? To address this question, I briefly introduced M-theory, which holds the universe has eleven dimensions, not simply the four dimensions (three spatial and one temporal) we typically encounter in our everyday lives. In effect, I suggested that the electron-positron pairs (the energy of a vacuum) might be in the hidden dimensions of M-theory.

In this chapter, we are going to view M-theory in a larger context, namely the context of a multiverse. The latest thinking in the multiverse world of science is that the multiverse consists of "branes," which collide to form universes. (note from Chapter 1, M-theory is shorthand for Membrane Theory.) While we can postulate that the collisions of branes form universes, does this explain the origin of the energy needed to create a universe? No! It just pushes the question back a level, namely: where do the branes come from?

We will start by understanding an overview of M-theory. First, as mentioned in Chapter 1, string theory is the foundation of M-theory. To understand this, we'll look at the development of these theories, their hypotheses, and the scientific problems they solve.

Most high school science classes teach the classical view of the atom, incorporating subatomic particles like protons, electrons, and neutrons. This is the particle theory of the atom dating to the early Twentieth Century. In about the 1960s, scientists discovered more subatomic particles. By the 1970s, scientists discovered that protons and neutrons consist of subatomic particles called quarks (an elementary particle not known to have a substructure). In the 1980s, a mathematical model called string theory, was developed. It is a branch of theoretical physics. String theory sought to explain how to construct all particles and energy in the universe via hypothetical one-dimensional "strings." Subatomic particles are no longer extremely small masses.

Instead, they are oscillating lines of energy, hence the name "strings." In addition, the latest string theory (M-theory) asserts that the universe is eleven dimensions, not the four-spacetime dimensions we currently experience in our daily lives. String theory was one of science's first attempts at a theory of everything (a complete mathematical model that describes all fundamental forces and matter).

In about the mid-1990s, scientists considered the equivalences of the various string theories. In the mid-1990s, the five leading string theories were combined into a one comprehensive theory, M-theory. M-theory postulates eleven dimensions of space filled with membranes, existing in the Bulk (super-universe). The Bulk contains an infinite number of membranes, or "branes" for short.

According to M-theory, when two branes collide, they form a universe. The collision is what we observed as the Big Bang when our universe formed. From that standpoint, universes continually form via other Big Bangs (collisions of branes).

Does this explain the true origin of the energy? No! It still begs the question: where does the energy come from to create the membranes? The even-bigger question: is there any scientific proof of the multiverse? As we discussed in Chapter 3, several scientists claim unusual ring patterns on the cosmic microwave background might be the result of other universes colliding with ours. However, even the scientists forwarding this theory suggest caution. It is speculative. At this point, we must admit no conclusive evidence of a multiverse exists. In fact, numerous problems with the multiverse theories are known. The next chapter will discuss them. This does not mean there are no multiverses. Currently, though, we have no conclusive experimental proof, but do have numerous unanswered questions.

"Absence of proof is not proof of absence."

Poet, William Cowper (1731-1800)

CHAPTER 6

A Multiverse of Problems

All multiverse theories share three significant problems.

1) None of the multiverse theories explains the origin of the initial energy to form the universe. They, in effect, sidestep the question entirely. Mainstream science believes, via inference, in the reality of energy. Therefore, it is a valid question to ask: what is the origin of energy needed to form a multiverse? In Chapter 2 ("Big Bang – Singularity or Duality?"), we discussed the theory of a "quantum fluctuation" as a candidate to explain the spontaneous creation of matter-antimatter energy, but this pushes the problem back one step. Some scientists will argue the energy came from dimensions we cannot experience.

This pushes the problem back yet another step. How did the energy get into those dimensions?

2) No conclusive experimental evidence proves that multiverses exist. This is not to say that they do not exist. Eventually, novel experiments may prove their existence. As discussed in a previous chapter, one group of such experiments involves super colliders. Super colliders force collisions between particles traveling near the speed of light. After the collision, scientists are trying to determine if any matter or energy is missing. If they are able to prove conclusively that there is missing matter or energy, this may support the existence of dimensions beyond the four spacetime dimensions we experience. To date, no conclusive data exists. Other methods of proof may emerge. Perhaps new types of telescopes not available today will enable us to look beyond the cosmic microwave background and "see" other universes. The critics of the multiverse have a point. To date, all we have are concepts and mathematical models. Critics assert, by its nature, the multiverse is not provable. Thus, we should not consider it science, but rather metaphysics. For the multiverse to step into the realm of science fact, we need valid scientific evidence. In a word, we need "proof."

3) Critics argue it is poor science. We are postulating universes we cannot see or measure in order to

> explain the universe we can see and measure. This is another way of saying it violates Occam's razor. This brings us back to the whole issue of proof. We have no evidence to the contrary that we have more than one universe.

In the last hundred years, we have made discoveries, and experimentally verified phenomena that in prior centuries would have been considered science fiction, metaphysics, magic, and unbelievable. We discovered numerous secrets of the universe, once believed to be only the Milky Way galaxy—to now being an uncountable number of galaxies in a space that is expanding exponentially. We also unlocked the secrets of the atom, once believed to be the fundamental building block of matter (from the Greek atomos "uncut"). Currently, we understand the atom consists of electrons, protons, and neutrons, which themselves consist of subatomic particles like quarks. The list of discoveries that have transformed our understanding of reality over the last century is endless. From my perspective, skepticism can be healthy. However, one cannot be entirely closed-minded when it comes to exploring the boundaries of science. As the respected physicist, Michio Kaku said in Nina L. Diamond's *Voices of Truth* (2000), "The strength and weakness of physicists is that we believe in what we can measure. And if we can't measure it, then we say it probably doesn't exist. And that closes us off to an enormous amount of phenomena that we may not be able to measure ..."

For those that like proof or at least evidence that plausibly could lead to proof, I am going to discuss one last "multiverse." We can plausibly "prove" it exists. Unfortunately, if we are in that multiverse, we do not exist.

"The human brain has..." [the ability to perform] "...20 million billion calculations per second. ... by the year 2020, (a massively parallel neural net computer) will have doubled about 23 times (from 1997's $2,000 modestly parallel computer that could perform around 2 billion connection calculations per second) ... resulting in a speed of about 20 million billion neural connection calculations per second, which is equal to the human brain."

Ray Kurzweil, The Age of Spiritual Machines, 1999

CHAPTER 7

Trapped in a Self-Conscious Supercomputer

Two words: Artificial Intelligence. Most people have heard about it. Perhaps you have read science-fiction books or seen science-fiction movies about it. What is it in the ideal fictional case? A computer that is able to learn and adapt on its own.

If it becomes self-aware, it can legitimately be considered another life form or even another universe.

Science fiction? No! Look at real-life results from the last 15 years.

In 1997, IBM's chess-playing computer "Deep Blue" became the first computer to beat world-class chess champion, Garry Kasparov. In a six-game match, Deep Blue prevailed by two wins to one with three draws. Until this point, no computer was able to beat a chess grandmaster. This garnered headlines worldwide, and was a milestone that embedded the reality of artificial intelligence into the consciousness of the average person.

In 2005, a robot conceived and developed at California's Stanford University, was able to drive autonomously for 131 miles along an unrehearsed desert trail, winning the DARPA Grand Challenge (the government's Defense Advanced Research Projects Agency prize for a driverless vehicle).

In 2007, Carnegie Mellon University's self-driving SUV called Boss made history by swiftly and safely driving 55 miles in an urban setting while sharing the road with human drivers. It, too, won the DARPA Urban Challenge.

In 2011, on an exhibition match on the popular TV quiz show, Jeopardy! , IBM's computer "Watson," defeated both of Jeopardy! greatest champions, Brad Rutter and Ken Jennings.

Today, we take artificial intelligence (AI) for granted. For example, computers and even smart phones have sophisticated chess-playing software. AI is part of the Xbox 360's algorithms for games. However, have we reached the point where a computer replicates a human mind? Not yet. One test held as the "gold standard" for this is the Turing test, proposed in 1950 by Alan Turing, an English mathematician, logician, cryptanalyst, and computer scientist. Turing is widely acknowledged as the father of computer science and artificial intelligence. In fact, Turing developed an electromechanical machine during WWII that helped break the German Enigma machine's code. The Turing test, which a computer must pass to demonstrate the computer replicates the human mind. The test requires that a machine (for example, a computer with voice synthesis) carry on a conversation with a human, and that other humans are able to hear the conversation (and not see the participants), and cannot distinguish the machine from the human.

Apple's Seri application for the iPhone is a small step in that direction. If you see Apple's TV commercials, people are talking to their phones, and phones are talking back. The conversations consist of the phone owners asking questions or giving simple commands to their iPhones. The commercial makes it appear that the iPhone passes the Turing test, but in reality, the conversations are limited to simple questions and simple commands. However, imagine what conversations with the iPhone will be like in about 20 years. The iPhone, and smart phones like it, will almost certainly pass the Turing test.

How close are we to a true artificial life form (similar to Lt. Commander Data in Star Trek: The Next Generation)? Most scientists believe we are extremely close. In fact, Ray Kurzweil (American author, scientist, inventor and futurist) has used Moore's law to calculate that desktop computers will be equivalent to human brains by the year 2029. Moore's law states the number of transistors that can be placed inexpensively on an integrated circuit doubles approximately every two years. By 2045, Kurzweil predicts, artificial intelligence will be able to improve itself faster than anything we can conceive. If this is true, by the mid Twenty-First Century, we may appear no smarter than insects to those machines. This is sometimes the theme of "how-will-the-world-end" type of documentaries, science-fiction books and movies. This is the whole premise behind the popular Terminator movies.

Now, we will return to our main point of a supercomputer universe. If indeed, computers one day will replicate a human mind, one can postulate that with time, it can replicate millions and eventually billions of such minds, each with its own self-awareness and personality. The minds inside the "machine" think they are real, and are in a universe. As more time passes, the machine can create another "universe." This scenario can continue forever, or until an unknown entity pulls the plug.

Could we be those people (minds inside a computer)? If you have a religious belief in a supreme being, in effect, we are those people in God's computer. If you do not hold religious beliefs, we could be those people in a race of advanced aliens' computer. In this scenario, a supernatural being or technology-advanced aliens gave the command to begin our existence. The command was simply, "Let there be light," and the super-computer program, simulating our existence and reality, began to run. If this is true, do we exist? The answer to that question depends on your viewpoint. We do not exist in the way we think we exist. We are all part of a sophisticated computer program in a supercomputer. If this is our reality, we are trapped in a supercomputer capable of replicating human minds, and imposing the construct of a universe on those minds.

At this point, I am going back to Occam's razor. With that as my guiding premise, I postulate our universe is real (exactly the way we experience it), we are real, and this book is real.

In chapters 1 through 7, we discussed several theories to address the ultimate question: what happened before the Big Bang? In chapter 8, we will use the information to approach an answer. Obviously, there is an answer. Our existence proves that. Will we end up with the right answer? That is a hard question in itself to answer, but it is important to try. As Stephen Hawking said, "We are just an advanced breed of monkeys on a minor planet of a very average star. But we can understand the Universe. That makes us something very special." Therefore, fellow monkeys, we must do our best to approach the right answer.

"We occasionally stumble over the truth but most of us pick ourselves up and hurry off as if nothing had happened."

Winston Churchill, 1874-1965

CHAPTER 8

Answering the Ultimate Question

The last seven chapters presented scientific theories acquired over centuries. Their purpose was to lay the foundation to answer the ultimate question: what happened before the Big Bang? The answer is incredibly difficult. We have no ability to access information prior to the Big Bang. In 2010 and 2011, scientific information regarding concentric rings on the cosmic microwave background began to fuel theories of other universes bumping into ours. This may eventually provide information about the universe prior to the Big Bang. However, this is a highly speculative theory at this point. Other phenomena may account for the concentric rings. We have no scientific consensus. Factually, we have no conclusive experimental evidence that dates prior to the Big Bang. Particle accelerators are unable to reproduce the energy levels associated with the Big Bang. Therefore, we do not have adequate information with which to formulate a verifiable theory. As a result, the scientific community has conflicting theories. Numerous paradigms—religious, scientific, and political—are converging to bend and shape the answer. As much as we try to be objective, we are all a product of our culture. We have serious problems to overcome. To approach an answer, we will take a systematic approach.

We will deal with the religious paradigms in a later chapter. For that reason, I suggest we take God out the equation in this chapter. In addition, I suggest we set politics aside. Although, I recognize politics has had profound effects on science, I am concerned that including a discussion on the political forces within the scientific community, the religious community, and government, will not help us arrive at an answer. However, I fully expect that any answer to the ultimate question will be debated, and take on political nuances. I believe our best chance at approaching an answer is to stick with the science. In truth, though, this will be subjective because we are dealing with the edge of scientific knowledge. We do not have a provable scientific answer to the question. We have our intellect, our imagination, and scientific observations, and will need all three elements to approach an answer. However, our imagination may ultimately play the largest role. As Einstein said, "I believe in intuition and inspiration. Imagination is more important than knowledge. For knowledge is limited, whereas imagination embraces the entire universe, stimulating progress, giving birth to evolution. It is, strictly speaking, a real factor in scientific research." With this in mind, we will begin.

The biggest clue we have with regard to the origin of energy for the Big Bang lies in the phenomena of virtual particles, as discussed in Chapter 1. Obviously, we have an experimentally verified phenomenon, namely that something can come from nothing. However, to a physicist, as we learned, nothing is something. Our laboratory experiments confirm a vacuum contains matter-antimatter pairs, which continually pop in and out of existence due to quantum fluctuations. Although, counterintuitive, we learned that a quantum fluctuation is a theory in quantum mechanics that argues, in accordance with the Heisenberg uncertainty principle, virtual particles can come from nothing, typically as matter-antimatter pairs.

The next big clue regarding the origin of the Big Bang's energy comes from Lawrence Krauss' bestselling book, *A Universe from Nothing: Why There Is Something Rather Than Nothing* (Free Press, 2012), which we discussed in Chapter 2. The main hypothesis forwarded by Dr. Krauss is that the Big Bang is the result of a quantum fluctuation, much like a virtual particle, except on a cosmic scale. This view still forwards the theory of a singularity, one infinitely dense particle. However, I argued that it was more likely the Big Bang was the result of a duality, namely the quantum fluctuation resulted in the spontaneous creation of an infinitely dense matter-antimatter particle pair. The Big Bang resulted when the pair collided, hence the name "duality." The reasoning for this argument came from our observation of virtual particles in the laboratory. We typically observe behavior that suggests the spontaneous creation of a matter-antimatter pair of virtual particles. I termed this theory The Big Bang Duality.

Next, in Chapter 3, we addressed the question: what caused the Big Bang to go bang? At this point, the Minimum Energy Principle was introduced.

Energy in any form seeks stability at the lowest energy state possible, and will not transition to a new state unless acted on by another energy source.

The Minimum Energy Principle is an abstraction of numerous similar laws, such as the second law of thermodynamics, held in various scientific contexts. It is consistent with the law of entropy, which is a measure of a system's disorderliness. Any increase in entropy implies a decrease in order. We will discuss entropy further a little later in this chapter.

If you accept the Minimum Energy Principle, it allows us to answer why the Big Bang went bang. The scientific community agrees that the Big Bang started with a point of infinite energy, the instant prior to the expansion. The infinite energy point is consistent with either the singularity theory, first proposed by Jesuit priest Georges LeMaitre in the late 1920s, or The Big Bang Duality theory, which I proposed. In either case, all logic urges us to conclude it was in an infinitely excited energy state. We have every reason to question its stability. As soon as the point of infinite energy came to exist, in accordance with the Minimum Energy Principle, it had to seek stability at a lower energy level. Therefore, the Big Bang was a form of energy dilution. The energy contained in the infinitely dense energy point sought a less-exited state, which requires the expansion of space associated with the Big Bang.

This has implications about entropy. In Chapter 3, we noted the Big Bang, and associated energy dilution, is consistent with the theory of entropy. The universe is moving to a less-orderly state. The entropy of the universe is increasing. This is counterintuitive, and most people view the universe as having a high degree of order. However, the formation of every star, planet, and galaxy, results in the universe's entropy increasing. In the formation of celestial bodies like stars and planets, heat emissions occur. The heat is no longer useful to do work. Therefore, the entropy increases. The increase in entropy is enormous, which implies the disorder in the universe is increasing enormously. Consider the formation of a single water molecule, when a hydrogen atom bonds with two oxygen atoms. The bonding process results in the spontaneous emission of heat. If we want to think about this on a larger scale, the infamous 1937 97-passenger German Hindenburg airship disaster is an example of the amount of heat emitted when hydrogen bonds with oxygen atoms. How much does entropy increase? Roughly, one burning match gives off about 1000 joules of energy, which is equal to about five million million million million units of entropy increase. The amount of entropy increase of the universe (the increase in disorder) as it goes through its daily routine—stars emitting energy, planets orbiting stars, the creation of new stars, and all the other celestial phenomena we observe—is incalculable. However, it is happening during every second of our lives. As a result, the disorder in the universe is enormously increasing. Eventually, the entropy of the universe will increase to the point that all energy becomes useless heat. We will reach the "entropy apocalypse."

Let's continue to address the ultimate question. We have summarized how the infinite energy point came to exist. Continuing with our summary, we asked how long the infinite energy point existed before it went "bang." Our observation of matter-antimatter collisions in the laboratory argues that the exact instant the infinite energy point came to exist, via the collision between the infinitely energy-dense matter-antimatter particles, it went bang. In a later chapter, we will discuss matter-antimatter collisions. At this point, it is essential to understand, from laboratory evidence, that when matter and antimatter collide, they immediately form a miniature Big Bang, along the lines described previously. This is inferential evidence that the Big Bang Duality theory may indeed deserve consideration. As stated earlier, the Big Bang Duality theory postulates the spontaneous creation of an infinitely energy-dense matter-antimatter pair in the Bulk. When they collide, the result is a Big Bang.

The concept that the Big Bang was the result of a quantum fluctuation strongly suggests that the scientific laws of our universe are the scientific laws of the Bulk. Natural laws, not our scientific theories and equations, but the actual laws that reality obeys, originate in the Bulk. This implies the physical laws of our universe pre-date the Big Bang, and it suggests that time pre-dates the Big Bang. The energy changes in the Bulk give meaning to time. Therefore, a cogent argument exists that time pre-dates the Big Bang.

Lastly, we will discuss how the multiverse theories fit into the above framework. In Chapter 2, we discussed the concept of a super-universe (the "Bulk"). If our universe's Big Bang is the result of a quantum fluctuation in the Bulk, we have reason to believe other Big Bangs are occurring in the Bulk. Multiple Big Bangs would imply numerous universes. This gives credence to elements of the multiverse theories (Chapters 4 -7).

Obviously, this summarizes the essence of the previous seven chapters, which are the distillation of centuries of scientific knowledge. For me, it provides a compelling picture of what occurred prior to the Big Bang. I ask that you consider it, and draw your own conclusions. At this point, we are on the edge of science, and proof is scarce. Elements of the above theories will eventually be verifiable in the laboratory as super colliders continue to provide insight into high-energy physics and the properties of antimatter. It is possible that the day will come when all the missing pieces of the jigsaw puzzle fall into place, and the entire picture reveals itself. For the present, we need to use our imaginations, and the scientific knowledge we have.

The universe we live in is wondrous and mysterious. A large number of mysteries still baffle science. If you love a compelling mystery, you will likely love Section II: What Mysteries Still Baffle Modern Science?

Section II

What Mysteries Still Baffle Modern Science?

"The most beautiful thing we can experience is the mysterious. It is the source of all true art and science."

—Albert Einstein (1879-1955)

"At the beginning, equal amounts of matter and antimatter were created (in the Big Bang). Now there seems to be only matter. There have been theoretical speculations about the disappearance of antimatter, but no experimental support."

Samuel Ting, American physicist who received the Nobel Prize in 1976, and a leading advocate in the search for antimatter in space

CHAPTER 9

Where Is the Missing Antimatter?

One of the great mysteries of our universe, and a weakness of the Big Bang theory, is that matter, not antimatter, totally makes up our universe. According to the Big Bang theory, there should be equal amounts of matter and antimatter. If there were any quantities of antimatter in our galaxy, we would see radiation emitted as it interacted with matter. We do not observe this. It is natural to ask the question: where is the missing antimatter? (Recall, that antimatter is the mirror image of matter. For example, if we consider an electron matter, the positron is antimatter. The positron has the same mass and structure as an electron, but the opposite charge. The electron has a negative charge, and the positron has a positive charge. Antimatter bears no relationship to dark matter. (Dark matter is discussed in the next chapter.)

Several theories float within the scientific community to resolve the missing antimatter issue. The currently favored theories (baryogenesis theories) employ sub-disciplines of physics and statistics to describe possible mechanisms. The baryogenesis theories start out with the same premise, namely the early universe had both baryons (an elementary particle made up of three quarks) and antibaryons (the mirror image of the baryons). At this point, the universe underwent baryogenesis. Baryogenesis is a generic term for theoretical physical processes that produce an asymmetry (inequality) between matter and antimatter. The asymmetry, per the baryogenesis theories, resulted in significant amounts of residual matter, as opposed to antimatter. The major differences between the various baryogenesis theories are in the details of the interactions between elementary particles. Baryogenesis essentially boils down to the creation of more matter than antimatter. In other words, it requires the physical laws of the universe to become asymmetrical. We need to understand what this means.

The symmetry of physical laws is widely accepted by the scientific community. What does "symmetry" mean in this context?

- First, it means that the physical laws do not change with time. If a physical law is valid today, it continues to be valid tomorrow, and any time in the future. This is a way of saying that a time translation of a physical law will not affect its validity.
- Second, it means that the physical laws do not change with distance. If the physical law is valid on one side of the room, it is valid on the other side of the room. Therefore, any space translation of a physical law will not affect its validity.
- Lastly, it means that the physical laws do not change with rotation. For example, the gravitational attraction between two masses does not change when the masses rotate in space, as long as the distance between them remains fixed. Therefore, any rotational translation of a physical law will not affect its validity.

This is what we mean by the symmetry of physical laws.

Next, we will address the asymmetry of physical laws. In this context, "asymmetry" means that the symmetry of physical laws no longer applies. For example, a law of physics may be valid in a specific location, but not in another, when both locations are equivalent. Is this possible? Maybe. There has been experimental evidence that the asymmetry is possible (a violation of the fundamental symmetry of physical laws). For example, radioactive decay and high-energy particle accelerators have provided evidence that asymmetry is possible. However, the evidence is far from conclusive. Most importantly, it does not fully explain the magnitude of the resulting matter of the universe.

This casts serious doubt on the baryogenesis theories. In addition, the baryogenesis theories appear biased by our knowledge of the outcome. By making certain (questionable) assumptions, and using various scientific disciplines, they result in the answer we already know to be true. The universe consists of matter, not antimatter. Therefore, baryogenesis theories may not be an objective explanation. However, apart from the Big Bang Duality theory, it is science's best theory of the missing antimatter dilemma.

The Big Bang Duality theory, suggested in Chapter 2, provides a simpler explanation, which does not violate the fundamental symmetry of physical laws. From this viewpoint, it deserves consideration.

In essence, the Big Bang Duality theory hypothesizes that the Big Bang was the result of a collision of two infinitely dense matter-antimatter particles in the Bulk. This theory rests on the significant experimental evidence that when virtual particles emerge from "nothing," they are typically created in matter-antimatter pairs (Chapter 1). Based on this evidence, I argued in Chapter 2, the Big Bang was a result of a duality, not a singularity as is often assumed in the Big Bang model. The duality would suggest two infinitely dense energy particles pop into existence in the Bulk. These are infinitely energy-dense "virtual particles." One particle would be matter, the other antimatter. The collision between the two particles results in the Big Bang.

What does this imply? It implies that the Big Bang was the result of a matter-antimatter collision. What do we know about those types of collisions from our experiments in the laboratory? Generally, when matter and antimatter collide in the laboratory, we get "annihilation." However, the laws of physics require the conservation of energy. Therefore, we end up with something, rather than nothing. The something can be photons, matter, or antimatter.

You may be tempted to consider the Big Bang Duality theory a slightly different flavor baryogenesis theory. However, the significant difference rests on the reactants, those substances undergoing the physical reaction, when the infinitely energy-dense matter-antimatter particles collide. The Big Bang Duality postulates the reactants are two particles (one infinitely energy-dense matter particle and one infinitely energy-dense antimatter particle). When the two particles collide, the laboratory evidence suggests the products that result are matter, photons, and antimatter. Contrary to popular belief, we do not get annihilation (nothing), when they collide. This would violate the conservation of energy. Consider this result. Two of the three outcomes, involving the collision of matter with antimatter, favor our current universe, namely photons and matter. In addition, conclusions drawn from the analysis of eight years of data from the Tevatron collider at the Department of Energy's Fermi National Accelerator resulted in a 2010 announcement that collisions between matter and antimatter favor the creation matter. This suggests that the collision of two infinitely dense matter-antimatter pairs statistically favor resulting in a universe filled with matter and photons. In other words, the universe we have. While not conclusive, it is consistent with the Big Bang being a duality. It is consistent with the reality of our current universe, and addresses the issue: where is the missing antimatter? The answer: The infinitely energy-dense matter-antimatter pair collides. The products of the collision favor matter and energy. Any resulting antimatter would immediately interact with the matter and energy. This reaction would continue until all that remains is matter and photons. In fact, a prediction of the Big Bang Duality theory would be the absence of observable antimatter in the universe. As you visualize this, consider that the infinitely energy-dense matter and antimatter particles are infinitesimally small, even to the point of potentially being dimensionless. Therefore, the collision of the two particles results in every quanta of energy in each particle contacting simultaneously.

You may be inclined to believe a similar process could occur from a Big Bang singularity that produces equal amounts of matter and antimatter. The problem with this theory is that the initial inflation of the energy (matter and antimatter) would quickly separate matter and antimatter. While collisions and annihilations would occur, we should still see regions of antimatter in the universe due to the initial inflation and subsequent separation. If there were such regions, we would see radiation resulting from the annihilations of antimatter with matter. We see only scant evidence of radiation in the universe that would suggest the existence of antimatter. Therefore, the scientific community has high confidence that the universe we observe consists almost entirely of matter.

I have sidestepped the conventional baryogenesis statistical analysis used to explain the absence of antimatter, which is held by most of the scientific community. However, the current statistical treatments require a violation of the fundamental symmetry of physical laws. Essentially, they argue the initial expansion of the infinitely dense energy point (singularity) produces more matter than antimatter, hence the asymmetry. This appears to complicate the interpretation, and violate Occam's razor. The Big Bang Duality theory preserves the conservation of energy law, and does not require a violation of the fundamental symmetry of physical laws.

Let me propose a sanity check. How comfortable is your mind (judgment) in assuming a violation of the fundamental symmetry of physical laws? I suspect many of my readers and numerous scientists may feel uncomfortable about this assumption. If you start with the Big Bang Duality theory, it removes this counterintuitive assumption. This results in a more straightforward, intellectually satisfying, approach, consistent with all known physical laws. Therefore, this theory fits Occam's razor.

Next, we will discuss another mystery of science, namely dark matter. It is one of the cosmos' greatest mysteries, and plays a key role in our survival.

"A cosmic mystery of immense proportions, once seemingly on the verge of solution, has deepened and left astronomers and astrophysicists more baffled than ever. The crux ... is that the vast majority of the mass of the universe seems to be missing."

William J. Broad, In "If Theory is Right, Most of Universe is Still "Missing."' New York Times (11 Sep 1984)

CHAPTER 10

The Mysterious Dark Matter

It was a dark and stormy night. We were hiding. All around us was dark matter, but we could not see it. We could feel its presence. We were trapped, and unable to move. The dark matter held us captive.

Okay, I know I started out this chapter on a melodramatic note. However, every word is true, figuratively speaking. Are you skeptical? This is one time I suggest you drop your defensive shields. Dark matter is real, mysterious, and necessary for our existence. Without it, we would not have a universe. It is a good thing with an ominous-sounding name. So, what is dark matter?

The most popular theory of dark matter is that it is a slow-moving particle. It travels up to a tenth of the speed of light. It neither emits nor scatters light. In other words, it is invisible. However, its effects are detectable, as I will explain below. Scientists call the mass associated with dark matter a "WIMP" (Weakly Interacting Massive Particle).

In 1933, Fritz Zwicky (California Institute of Technology) made a crucial observation. He discovered the orbital velocities of galaxies were not following Newton's law of gravitation (every mass in the universe attracts every other mass with a force inversely proportional to the square of the difference between them). They were orbiting too fast for the visible mass to be held together by gravity. If the galaxies followed Newton's law of gravity, the outermost stars would be thrown into space. He reasoned there had to be more mass than the eye could see, essentially an unknown and invisible form of mass that was allowing gravity to hold the galaxies together. Zwicky's calculations revealed that there had to be 400 times more mass in the galaxy clusters than what was visible. This is the mysterious "missing-mass problem." It is normal to think that this discovery would turn the scientific world on its ear. However, as profound as the discovery turned out to be, progress in understanding the missing mass lags until the 1970s.

In 1975, Vera Rubin and fellow staff member Kent Ford, astronomers at the Department of Terrestrial Magnetism at the Carnegie Institution of Washington, presented findings that reenergized Zwicky's earlier claim of missing matter. At a meeting of the American Astronomical Society, they announced the finding that most stars in spiral galaxies orbit at roughly the same speed. They made this discovery using a new, sensitive spectrograph (a device that separates an incoming wave into a frequency spectrum). The new spectrograph accurately measured the velocity curve of spiral galaxies. Like Zwicky, they found the spiral velocity of the galaxies was too fast to hold all the stars in place. Using Newton's law of gravity, the galaxies should be flying apart, but they were not. Presented with this new evidence, the scientific community finally took notice. Their first reaction was to call into question the findings, essentially casting doubt on what Rubin and Ford reported. This is a common and appropriate reaction, until the amount of evidence (typically independent verification) becomes convincing.

In 1980, Rubin and her colleagues published their findings (V. Rubin, N. Thonnard, W. K. Ford, Jr, (1980). "Rotational Properties of 21 Sc Galaxies with a Large Range of Luminosities and Radii from NGC 4605 (R=4kpc) to UGC 2885 (R=122kpc)." Astrophysical Journal 238: 471.). It implied that either Newton's laws do not apply, or that more than 50% of the mass of galaxies is invisible. Although skepticism abounded, eventually other astronomers confirmed their findings. The experimental evidence had become convincing. "Dark matter," the invisible mass, dominates most galaxies. Even in the face of conflicting theories that attempt to explain the phenomena observed by Zwicky and Rubin, most scientists believe dark matter is real. None of the conflicting theories (which typically attempted to modify how gravity behaved on the cosmic scale) was able to explain all the observed evidence, especially gravitational lensing (the way gravity bends light).

Currently, the scientific community believes that dark matter is real and abundant, making up as much as 90% of the mass of the universe. However, dark matter is still a mystery. For years, scientists have been working to find the WIMP particle to confirm dark matter's existence. All efforts have been either unsuccessful or inconclusive.

The Department of Energy Fermi National Accelerator Laboratory Cryogenic Dark Matter Search (CDMS) experiment is ongoing, in an abandoned iron mine about a half mile below the surface, in Soudan, Minnesota. The Fermilab is a half mile under the earth's surface to filter cosmic rays so the instruments are able to detect elementary particles without the background noise of cosmic rays. In 2009, they reported detecting two events that have characteristics consistent with the particles that physicists believe make up dark matter. They may have detected the WIMP particle. However, they are not making that claim at the time of this writing. The Fermilab stopped short of claiming they had detected dark matter because of the strict criteria that they have self-imposed, specifically there must be less than one chance in a thousand that the event detected was due to a background particle. The two events, although consistent with the detection of dark matter, do not pass that test.

From an article written in Fermilab Today (December 13,

2009), the Fermilab Director Pier Oddone said, "While this result is consistent with dark matter, it is also consistent with backgrounds. In 2010, the collaboration is installing an upgraded detector (SuperCDMS) at Soudan with three times the mass and lower backgrounds than the present detectors. If these two events are indeed a dark matter signal, then the upgraded detector will be able to tell us definitively that we have found a dark matter particle." As of this writing, Fermilab and other laboratories maintain their quest to find the WIMP particle. To date, we are without conclusive evidence that the WIMP exists.

If it exists, there is a reasonable probability that the WIMP particle can be "created" via experiments involving super colliders (such as the Large Hadron Collider (LHC) built by the European Organization for Nuclear Research (CERN) over a ten-year period from 1998 to 2008). Super colliders have successfully given us a glimpse into the early universe. Since most scientists believe that dark matter exists as part of creation at the instant of the Big Bang, super colliders may provide a reasonable methodology of directly creating dark matter. As of this writing, scientists using the Large Hadron Collider are attempting to create WIMP particles via high-energy proton collisions.

Are we on the right track? Is there a WIMP particle? Consider the experimental evidence.

The rotation of stars, planets, and other celestial masses orbit galaxies, like ours, too rapidly relative to their mass and the gravitational pull exerted on them in the galaxy. For example, an outermost star should be orbiting slower than a similar-size star closer to the center of the galaxy, but we observe they are orbiting at the same rate. This means they are not obeying Newton's laws of motion or Einstein's general theory of relativity. This faster orbit of the outermost stars suggests more mass is associated with the stars than we are able to see. If not, the stars would fly free of their orbits, into outer space.

We can see the effect dark matter has on light. It will bend light the same way ordinary matter bends light. This effect is gravitational lensing. The visible mass is insufficient to account for the gravitational lensing effects we observe. Once again, this suggests more mass than what we can see.

We are able to use the phenomena of gravitational lensing to determine where the missing mass (dark matter) is, and we find it is throughout galaxies. It is as though each galaxy in our universe has an aura of dark matter associated with it. We do not find any dark matter between galaxies.

Most of the scientific community accepts the experimental evidence confirming the existence of dark matter. The real issue appears centered on the WIMP particle's existence.

Does the WIMP particle exist? Consider the facts.

1) The Standard Model of particle physics does not predict a WIMP particle. The Standard Model, refined to its current formulation in the mid-1970s, is one of science's greatest theories. It successfully predicted bottom and top quarks prior to their experimental confirmation in 1977 and 1995, respectively. It predicted the tau neutrino prior to its experimental

confirmation in 2000, and the Higgs boson prior to its experimental confirmation in 2012. Modern science holds the Standard Model in such high regard that a number of scientists believe it is a candidate for the theory of everything. Therefore, it is not a little "hiccup" when the Standard Model does not predict the existence of a particle. It is significant, and it might mean that the particle does not exist.

2) All experiments to detect the WIMP particle have to date been unsuccessful, including considerable effort by Stanford University, University of Minnesota, and Fermilab.

That is all the experimental evidence we have. Where does this leave us? The evidence is telling us the WIMP particle might not exist. We have spent about ten years, and unknown millions of dollars, which so far leads to a dead end. This appears to beg a new approach.

To kick off the new approach, consider the hypothesis that dark matter is a new form of energy. We know from Einstein's mass-energy equivalence equation ($E = mc^2$), that mass always implies energy, and energy always implies mass. For example, photons are massless energy particles. Yet, gravitational fields influence them, even though they have no mass. That is because they have energy, and energy, in effect, acts as a virtual mass.

If dark matter is energy, where is it and what is it? Consider these properties of dark-matter energy:

- It is not in the visible spectrum, or we would see it.
- It does not strongly interact with other forms of energy or matter.
- It does exhibit gravitational effects, but does not absorb or emit electromagnetic radiation.

Based on these properties, we should consider M-theory (the unification of superstring theories discussed in Chapter 5). Several prominent physicists, including one of the founders of string theory, Michio Kaku, suggest there may be a solution to M-theory that quantitatively describes dark matter and cosmic inflation. If M-theory can yield a superstring solution, it would go a long way to solving the dark-matter mystery. I know this is like the familiar cartoon of a scientist solving an equation where the caption reads, "then a miracle happens." However, it is not quite that grim. What I am suggesting is a new line of research and theoretical enquiry. I think the theoretical understanding of dark matter lies in M-theory. The empirical understanding lies in missing-matter experiments.

What is a missing-matter experiment? Scientists are performing missing-matter experiments as I write this book. They involve high-energy particle collisions. By accelerating particles close to the speed of light, and causing particle collisions at those speeds, they account for all the energy and mass pre- and post-collision. If any energy or mass is missing post-collision, the assumption would be it is in one of non-spatial dimensions predicted by M-theory.

Why would this work? M-theory has the potential to give us a theoretical model of dark matter, which we do not have now. Postulating we are dealing with energy, and not particles, would explain why we have not found the WIMP particle. It would also explain why the Standard Model of particle physics doesn't predict a WIMP particle. Postulating that the energy resides in the non-spatial dimensions of M-theory would explain why we cannot see or detect it. Real-world phenomena take place in the typical three spatial dimensions and one temporal dimension. If dark matter is in a different dimension, it cannot interact with "real"-world phenomena, except to exhibit gravity. Why is dark matter able to exhibit gravity? That is still a mystery, as is gravity itself. We have not been able to find the "graviton," the mysterious particle of gravity that numerous particle physicists believe exists. Yet, we know gravity is real. It is theoretically possible that dark matter (perhaps a new form of energy) and gravity (another form of energy) are both in a different dimension. This framework provides an experimental path to verify both the M-theory, and the existence of dark matter (via high-energy particle collisions).

This is a conceptual framework, but fits the observations. I am not suggesting we abandon our search for the WIMP particle. However, I suggest we widen our search to include the possibility that dark matter is not a particle, but a new form of energy.

Are you ready for another dark mystery? This one is even scary. The phenomenon is real, and spells doom for the universe. I am talking about dark energy. Is dark energy real? Is it causing the expansion of the universe to accelerate? The fate of the universe is in the balance—so we will try to figure this out.

"The universe is made mostly of dark matter and dark energy, and we don't know what either of them is..."

Saul Perlmutter, American astrophysicist, who shared both the 2006 Shaw Prize in Astronomy and the 2011 Nobel Prize in Physics with Brian P. Schmidt and Adam Riess for providing evidence that the expansion of the universe is accelerating.

CHAPTER 11

Is Dark Energy Real or Simply a Scary Ghost Story?

If it is not real, it is an extremely scary ghost story. Unfortunately, the phenomena we call dark energy is real. If it plays out on its current course, we are going to be alone, all alone. The billions upon billions of other galaxies holding the promise of planets with life like ours will be gone. The universe will be much like what they taught our grandparents at the beginning of the Twentieth Century. It will consist of the Milky Way galaxy. All the other galaxies will have moved beyond our cosmological horizon, and be lost to us forever. There will be no evidence that the Big Bang ever occurred.

Mainstream science widely accepts the Big Bang as giving birth to our universe. Scientists knew from Hubble's discovery in 1929 that the universe was expanding. However, prior to 1998, scientific wisdom was that the expansion of the universe would gradually slow down, due to the force of gravity. We were so sure, so we decided to confirm our theory by measuring it. Can you imagine our reaction when our first measurement did not confirm our paradigm, namely that the expansion of the universe should be slowing down?

What happened in 1998? The High-z Supernova Search Team (an international cosmology collaboration) published a paper that shocked the scientific community. The paper was: Adam G. Riess et al. (Supernova Search Team) (1998). "Observational evidence from supernovae for an accelerating universe and a cosmological constant." Astronomical J. 116 (3). They reported that the universe was doing the unthinkable. The expansion of the universe was not slowing down—in fact, it was accelerating. Of course, this caused a significant ripple in the scientific community. Scientists went back to Einstein's general theory of relativity and resurrected the "cosmological constant," which Einstein had arbitrarily added to his equations to prove the universe was eternal and not expanding. Previous chapters noted that Einstein considered the cosmological constant his "greatest blunder" when Edwin Hubble, in 1929, proved the universe was expanding.

Through high school-level mathematical manipulation, scientists moved Einstein's cosmological constant from one side of the equation to the other. With this change, the cosmological constant no longer acted to keep expansion in balance to result in a static universe. In this new formulation, Einstein's "greatest blunder," the cosmological constant, mathematically models the acceleration of the universe. Mathematically this may work, and model the accelerated expansion of the universe. However, it does not give us insight into what is causing the expansion.

The one thing that you need to know is that almost all scientists hold the paradigm of "cause and effect." If it happens, something is causing it to happen. Things do not simply happen. They have a cause. That means every bubble in the ocean has a cause. It would be a fool's errand to attempt to find the cause for each bubble. Yet, I believe, as do almost all of my colleagues, each bubble has a cause. Therefore, it is perfectly reasonable to believe something is countering the force of gravity, and causing the expansion to accelerate. What is it? No one knows. Science calls it "dark energy."

That is the state of science as I write this book in the latter half of 2012. The universe's expansion is accelerating. No one knows why. Scientists reason there must be a cause countering the pull of gravity. They name that cause "dark energy." Scientists mathematically manipulate Einstein's self-admitted "greatest blunder," the "cosmological constant," to model the accelerated expansion of the universe.

Here is the scary part. In time, we will be entirely alone in the galaxy. The accelerated expansion of space will cause all other galaxies to move beyond our cosmological horizon. When this happens, our universe will consist of the Milky Way. The Milky Way galaxy will continue to exist, but as far out as our best telescopes will be able to observe, no other galaxies will be visible to us. What they taught our grandparents will have come true. The universe will be the Milky Way and nothing else. All evidence of the Big Bang will be gone. All evidence of dark energy will be gone. Space will grow colder, almost devoid of all heat, as the rest of the universe moves beyond our cosmological horizon. The entire Milky Way galaxy will grow cold. Our planet, if it still exists, will end in ice. How is that for a scary story?

It is a bleak picture. You may need good news at this point, so here's some. This will all take billions of years to happen. Therefore, do not quit your day job or run up large debts. The end is not around the corner.

Will this mean the end of the Earth? Yes, but another piece of distressing news might end our Earth before the accelerated expansion of the universe does. Our sun is about five billion years old and is halfway through its expected life. Therefore, in about five billion years, our sun will die, which will mean an end to the Earth. The sole chance the human race has to survive beyond five billion years is to leave the Earth before the sun dies. That will mean we will have to master space travel, and be able to reach another Earth-like planet or become nomads, wandering the galaxy in spaceships. Tuck this concept in your subconscious mind for a later chapter when we discuss Stephen Hawking's warning against contacting advanced aliens.

The data and projected outcome are irrefutable. Given the present course the universe is on, eventually the Milky Way galaxy will be utterly alone. Is all this due to dark energy? If so, what causes dark energy? This is a hotly debated scientific theory, with a number of scientists dismissing it, and other scientists developing new theories to explain it. Here's a small glimpse into the debate.

There are currently two principal schools of thought regarding the theory of dark energy. I already mentioned the "cosmological constant" group. The second is "quintessence."

The quintessence model attributes the universe's acceleration to a fifth fundamental force that changes over time. The quintessence school of thought has its own equation. It differs from the cosmological constant equation by allowing the equation itself to change over time. In brief, the cosmological constant is a constant, and does not vary with time. The quintessence equation varies with time.

There are numerous less-popular schools of thought regarding what is causing the accelerating universe. These include string theory explanations, M-theory explanations, and a novel theory that the accelerated expansion is an illusion. This last theory comes from Christos Tsagas, a cosmologist at Aristotle University of Thessaloniki in Greece. Dr. Tsagas argues that the accelerated expansion of the universe is an illusion. It is due, he asserts, to the motion of the Earth, within the Milky Way galaxy, relative to the rest of the universe. If Dr. Tsagas is right, we do not need dark energy—and our worries are over. However, as explained shortly, there is little chance that the accelerated expansion of the universe is an illusion.

To shed light on this subject, start with what the data tells us.

The universe's expansion is accelerating. A recent census of 200,000 galaxies up to 7 billion years old appears to confirm that dark energy is real, and is, indeed, responsible for the accelerated expansion of the universe.

- As reported in www.space.com, *Dark Energy Is Real, New Evidence Indicates*, May 19, 2011. The report contains data collected over a five-year period using NASA's Galaxy Evolution Explorer (GALEX) and the Anglo-Australian Telescope atop Siding Spring Mountain in Australia.
- Discovery News also covered this study and reported it on their Website (http://news.discovery.com/space/dark-energy-universe-einstein-110519.html). The Discovery News report added that Einstein's cosmological constant properly models the accelerated expansion of the universe.

According to Discovery News, "The paper to be published in the Monthly Notices of the Royal Astronomical Society has been put together by a team of 26 scientists including Dr. Chris Blake from Melbourne's Swinburne University. It provides the first independent confirmation of both the existence of dark energy and its rate of expansion." However, while the study confirms the accelerated expansion of the universe as modeled by Einstein's cosmological constant, the study does not provide insight into the exact physics of dark energy.

Because of the accelerating expansion, entire galaxies are moving away from us faster than the speed of light. The more distant the galaxy, the faster it is accelerating away from us.

However, the galaxies themselves are not expanding. This is a scientific fact. Our Milky Way galaxy is behaving exactly as we would expect, with no expansion of the space between stars within the galaxy. The question becomes why. Is space between stars equal to space between galaxies? No, it is not. The space between stars and other celestial bodies within our galaxy appears glued together with dark matter, discussed in the last chapter. Dark matter does not appear to exist between distant galaxies. Gravitational attraction exists between galaxies, but no dark matter connects one distant galaxy to another. Therefore, a galaxy acts more like a single mass in the vastness of the universe. This is fortunate for us, or we would be moving away from our sun, which would have dire consequences.

This is all we know. This is the observational data. What mathematical model is correct? Is it the cosmological-constant or quintessence model? The most recent evidence appears to support the cosmological-constant modeling. However, no study to date adequately explains the physics of dark energy.

As stated earlier, almost all scientists strongly hold the belief that every effect has a cause (the "cause-and-effect" paradigm). With the concept of cause and effect in mind, I am going to present three highly speculative theories to explain the physics behind the accelerated expansion of the universe. This speculation rests on questionable scientific reports, scientific reasoning, and a significant amount of imagination. Therefore, they will read like science fiction. I present them because they may serve a purpose. First, the theories themselves may have a kernel of truth, and therefore, merit consideration. Second, they may inspire others to refine them, or use them as a springboard to propose other theories. As Einstein pointed out, imagination is essential in solving scientific mysteries. The interesting thing about the three theories is that ultimately we can prove or disprove them. In the end, they do not require faith. Please remember, however, at this point, they are speculative theories.

The Fifth-Force Theory—This theory postulates that there is a fifth force, beyond the four fundamental forces—gravity, strong nuclear, weak nuclear, and electromagnetism. This new force is "repulsion." The science behind the fifth force has a checkered past. In theory, it is about the same strength as gravity. The range of the fifth force is unknown, but for it to be a true cause of the universe's accelerating expansion, it would need to have a range similar to gravity. Gravity has an infinite range. It can range from small distances, like a rock falling from your hand toward the Earth. It can range to large distances like the pull that the sun exerts on the Earth, in order to keep it in orbit. Everything in the universe pulls on everything else.

Between 1885 and 1909, Loránd Eötvös and his team proved with a high degree of accuracy that the inertial mass and gravitational mass were equivalent. This had long been suspected, dating back to Isaac Newton (1642–1727). In 1986, five researchers, Ephraim Fischbach, Daniel Sudarsky, Aaron Szafer, Carrick Talmadge, and S. H. Aronson, were reanalyzing Eötvös' results and reported finding a fifth force ("Reanalysis of the Eötvös experiment," Physical Review Letters 56 3, 1986). However, subsequent experiments by other researchers failed to duplicate their results and interest in the fifth force was slowly dying until 2011. On April 7, 2011, physicists at Fermilab's circular particle accelerator, Tevatron, announced the possible discovery of a new particle, which they believed could be evidence of the elusive fifth force. Unfortunately, DZero's worldwide collaboration of scientists analyzed data from trillions of particle collisions, and was unable to verify the result. Physicists at the Large Hadron Collider (LHC), the world's highest-energy particle collider, located beneath the Franco-Swiss border near Geneva, Switzerland, are on the watch for this new particle. To date, they have not reported finding it.

Obviously, we have enough data to pique our curiosity and suggest the fifth force's reality. The experimental results are inconclusive, but more experiments are in the works. The Large Hadron Collider is scheduled for an upgrade in 2018, after 10 years of operation, to enable greater energy and detection capabilities. The planned upgrades are still under consideration. Perhaps following these upgrades, we will get the insight needed to prove or disprove a fifth force.

The Existence Theory—In the next chapter related to the nature of time, I will present a theory that existence, which is movement in time, requires enormous negative energy. I will also derive an equation, named the "Existence Equation Conjecture," which models this behavior. Due to the enormous negative energy implied by the Existence Equation Conjecture, I will theorize that it is drawing the energy required for existence from the universe. In effect, I am hypothesizing that the energy required for existence gives rise to what science terms dark energy, and it is causing the accelerated expansion of space. This theory is speculative. However, a data point suggests the Existence Equation Conjecture, itself, is valid, and is able to predict the particle life of subatomic particles accelerated to near the speed of light. For example, Appendix II provides the Existence Equation Conjecture's prediction of the energy required to extend a muon's life, an elementary particle, by approximately a factor of 30. If we compare this prediction to experimental data, we find agreement within 2%. This is excellent agreement, and suggests the Existence Equation Conjecture may be correctly modeling reality. Further verifications are possible using data from particle colliders like the Large Hadron Collider. If true, it provides insight into the physical process behind time dilation and the accelerated expansion of the universe. We will move on for now, but discuss this more fully when we discuss "The Mysterious Relationship between Time, Existence, and Energy" in a later chapter.

The Minimum Energy Principle—The accelerated expansion of space might be a result of the Minimum Energy Principle: *Energy in any form seeks stability at the lowest energy state possible, and will not transition to a new state unless acted on by another energy source.* The Minimum

Energy Principle, discussed in Chapter 3, is a generalized restatement of similar laws found in the physical sciences, and is independent of the scientific context. It abstracts the essence of the contextual statements, and views applications of the law in various scientific contexts as specific cases.

How does this law apply to the expansion of the universe? It implies the universe is expanding to a lower-energy density state, and slowly the heat of the universe is becoming widely dispersed, which suggest entropy is increasing. We have little doubt that the universe is following the Minimum Energy Principle. The issue becomes: does this explain the fundamental physical process of an accelerating universe? The answer is "maybe."

The scenario goes something like this. The Minimum Energy Principle initiates the expansion of the universe, and requires the energy of the universe to seek the lowest energy state possible. The Minimum Energy Principle is independent of distance. Gravity, on the other hand, is inversely proportional to square of the distance between objects. Therefore, the effects of gravity become rapidly and significantly weaker as objects move farther apart. This may allow the energy of the universe, in the form of galaxies and other celestial bodies, to increase the rate that they seek the lowest energy state possible. In effect, the Minimum Energy Principle becomes dominant as galaxies move further apart, since the effects of gravity decrease.

Why did I use the word "maybe" to answer whether this explains the accelerated expansion of the universe? Because we do not fully understand the Minimum Energy Principle.

It is possible to argue that the Minimum Energy Principle is a natural law of our universe, but that does not explain the "why" behind it. Here is an example to illustrate dilemma. We know gravity will cause a rock, released from your hand, to fall to Earth. The rock falls to Earth, in accordance with the Minimum Energy Principle, to lower its gravitational potential energy. However, it does not tell us why the rock seeks to lower it gravitational potential energy. We observe the Minimum Energy Principle to be true, but do not totally understand why it is true.

We have not solved the mystery of dark energy, although we have surfaced potential candidates that may lead to a solution. We know that the universe's expansion is accelerating, confirmed by numerous independent measurements. However, we have not pinned down the physical process behind the accelerated expansion. Experiments to find a cause have been inconclusive, all serving to deepen the mystery. As of this writing in 2012, dark energy remains an elusive mystery. However, we will bring greater insight to the mystery when we discuss "The Mysterious Relationship between Time, Existence, and Energy" in the next chapter.

At first, our next topic may not seem like much of a mystery. In fact, you may even dismiss it as ridiculous. If you do, you will be ill prepared for the journey that I will invite you to take with me in a later chapter—a journey involving time travel.

Our next mystery provides a crucial clue to time travel. It has to do with a simple question. What is time? Try defining it without using the word "time." It is not easy. Try explaining it scientifically. It is even harder. Yet, in the next chapter, we will do that. When we are done, we will have a new insight into a unique relationship between time, existence, and energy. Interested? All it takes is time. Please read on.

"As long as no one asks me [what time is] then I know it, but if someone asks me to explain it, I don't know it."

Anonymous

CHAPTER 12

The Mysterious Relationship Between Time, Existence, and Energy

What time is it? You have probably asked and been asked that question countless times. It does not seem to be a difficult question. So why am I calling attention to a possible mystery involving time? Consider this remarkable fact. The true nature of time is elusive. In this chapter, we will explore what few, if any before us, have explored. When we have completed our exploration, we will have a new insight into a relationship between time, existence, and energy. We will be far out on the limb of speculation, but scientific evidence shows that we may be onto a new secret of the universe. However, save that thought for a little later in this chapter, while we continue with understanding the nature of time.

Einstein proved that space and time were inseparably weaved into a four-dimension coordinate system, space-time, and time is highly dependent on the speed and positions of the observers. However, in our everyday life, we typically treat time as an absolute, independent of space. Our everyday life tends to reinforce that time for you is the same as time for me, even if you are calling me from a jet plane traveling at six hundred miles per hour. The reason is that, in our everyday life, we do not move fast enough to observe relativistic space-time differences. Space does not typically become part of the equation when people ask us the time.

To understand how mysterious time is, we will start by asking a basic question. What is time? This is not a new question. Scientists and philosophers have been asking an endless stream of questions about the nature of time.

- What causes time?
- Why does time appear to slow down by gravity and motion?
- Is time real or an emergent concept, namely the present is real, and our minds create the illusion of the past, and anticipate a future?
- Does time have a direction? For example, a person is born before they can die, which philosophers refer to as the "arrow of time."
- Is time a dimension?

- How do we measure time?
- Is time simply a measure of change, such as sand flowing through an hourglass, or does time exist when nothing is changing?

Philosophers have debated the nature of time for over 2500 years, and have left us with three principal theories, listed below in no particular order

1) Presentists Theory of Time—The "presentists" philosophers argue that present objects and experiences are real. The past and future do not exist. This would argue that time is an emerging concept, and exists in our minds. If this philosophy of time is correct, it renders time travel meaningless.
2) Growing-Universe Theory of Time—The "growing-universe" philosophers argue that the past and present are real, but the future is not. Their reasoning is the future has not occurred. Therefore, they reason the future is indeterminate, and not real. If this philosophy of time is correct, time travel may only be possible to the past.
3) Eternalism Theory of Time—The "eternalism" philosophers believe that there are no significant differences among present, past, and future because the differences are purely subjective. Observers at vastly different distances from an event would observe it differently because the speed of light is finite and constant. The farthest-away observer may

> be seeing the birth of a star while the closest observer may be seeing the death of the same star. In effect, the closest observer is seeing what will be the future for the farthest-away observer. If this philosophy of time is correct, this appears to open the possibility of time travel to both the past and the future.

From the above, clearly only the growing-universe and eternalism philosophies allow even the possibility of time travel. I am pointing this out because in the next chapter we are going to discuss time travel. I thought it would be intriguing to understand how the different philosophies of time view time travel. Next, we will turn our attention to science's answer to the question: what is time?

Almost everyone agrees that time is a measure of change, for example, the ticking of a clock as the second hand sweeps around the dial represents change. If that is true, time is a measure of energy because energy is required to cause change. Numerous proponents of the "Big Bang" hold that the Big Bang itself gave birth to time. They argue that prior to the Big Bang, time did not exist. This concept fits well into our commonsense notion that time is a measure of change. However, what if no change occurs? Suppose we are able to freeze one atom to absolute zero. This is a thought experiment. The laws of thermodynamics state it is impossible for us to reach absolute zero. However, for our purposes, imagine for a moment that we could suspend the laws of thermodynamics, and freeze an atom at absolute zero. At absolute zero, the atom would be devoid of heat energy, and all motion in the atom itself would cease. Even the position of the electrons, neutrons, and protons would freeze in place, which would violate Heisenberg's uncertainty principle. Once again, this is a thought experiment, so we are allowed to suspend the laws of physics to make a point. In this hypothetical scenario, does time have any meaning for the atom? Does time still exist? Most scientists say no. Without change, time is meaningless. Yet, the atom exists, even if we were to violate the laws of physics, and freeze it at absolute zero. It would still have mass, and therefore, exist. Observers viewing the atom would continue to measure time, and even tell us how long the atom has been at absolute zero. This suggests that existence implies time, even in the absence of energy change. Therefore, time in this sense is not a measure of change for the atom, but rather a measure of existence. If the current observers leave, and the atom is rediscovered by new observers millions of years later, it still exists.

Our modern conception of time comes from Einstein's special theory of relativity. In this theory, the rates of time run differently, depending on the relative motion of observers, and their spatial relationship to the event under observation. For example, picture an observer O_1 holding a clock C_1 at rest, relative to another observer O_2, traveling at near the speed of light, while holding a clock C_2. When observer O_1 looks at his clock, it appears to be running normally, but when he looks at the clock (C_2), in the hand of observer O_2, it appears to be running slowly. Time dilation causes this effect, discussed later in this chapter. Einstein's special theory of relativity predicts this effect. To demonstrate time dilation, consider the illustration in Figure 1:

O_1 holding a clock C_1
(At rest)

O_2 holding a clock C_2
(Traveling near the speed of light)

Figure 1 – O1 observes his clock C1 running normally, but when he looks at the clock (C_2), being held by observer O_2, it appears to be running slowly.

Now, consider another example, which illustrates how space and the speed of light influence an observer's perception of time. Consider two observers. One observer (O_1) close to Event A (for example, the event can be a dying star), and another observer (O_2), about ten billion light years away. The observer (O_1) is close to the event, and witnesses changes almost instantaneously as they occur. Observer (O_2) is ten billion light years away, and witnesses events with a delay of ten billion years. Therefore, observer (O_1) may be witnessing the death of a star. Observer (O_2) is witnessing

Figure 2 – Observer 1, close to Event A is observing the death of a star, while Observer 2, ten billion light years away, is observing the birth of the same star.

These two examples illustrate the concept of space-time. According to this view of time, we live on a world line, defined as the unique path of an object as it travels through four-dimensional space-time, rather than a timeline. At this point, it is reasonable to ask: what is the fourth dimension?

The fourth dimension is often associated with Einstein,

and typically equated with time. However, it was German mathematician Hermann Minkowski (1864-1909), who enhanced the understanding of Einstein's special theory of relativity by introducing the concept of four-dimensional space, since then known as "Minkowski space-time."

In the special theory of relativity, Einstein used Minkowski's four dimensional space—X_1, X_2, X_3, X_4, where X_1, X_2, X_3 are the typical coordinates of the three dimensional space—and $X_4 = ict$, where $i = \sqrt{-1}$, c is the speed of light in empty space, and t is time, representing the numerical order of physical events measured with "clocks." (The mathematical expression i is an imaginary number because it is not possible to solve for the square root of a negative number.) Therefore, $X_4 = ict$, is a spatial coordinate, not a "temporal coordinate." This forms the basis for weaving space and time into space-time.

To understand the nature of time, we have to go a little deeper into the science surrounding time. In the last decades of the Nineteenth Century, scientific experiments proved light has the same speed, regardless of whether the source of light is moving toward an observer or is moving away from an observer. This means the speed of light is a constant when measured in any inertial frame (a frame of reference moving at a constant velocity). This experimental evidence turned the world of scientists upside down.

For hundreds of years prior to this discovery, scientists embraced the Galilean transformation to calculate coordinates between two inertial frames of reference. The Galilean transformation worked if the two frames were at rest, or were traveling at velocities significantly below the speed of light. Here is an example of the Galilean transformation of time between two events.

> Consider a train passing through a train station with a big clock on a tower over the station. Assume the station clock and your wristwatch had both been set to the correct time prior to boarding the train at another location, and that both keep accurate time. Assume that the train does not stop at this station, and will continue its current speed of 60 miles per hour. As the train passes the station, you look at your wristwatch and the clock on the station tower. They both read the same time. If we call time on the tower $t_{1,}$ and time on your wristwatch t_2, the Galilean transformation requires that $t_1 = t_2$. In other words, time is an absolute, independent of the movement of the train, and the space between your wristwatch and the tower. The Galilean transformation works well as long as the speed of the train remains relatively slow compared to the speed of light. However, the Galilean transformation falls apart as the speed is increased. For example, assume you are passing the train station in a jet plane. Due to the time dilation effect predicted by Einstein's special theory of relativity, an observer at the station would see your wristwatch moving slower than the tower clock. (Pretend the observer uses binoculars to see your wristwatch.)

Although the Galilean transformation was a cornerstone of Newtonian mechanics, which considered time a universal absolute, it proved inadequate to describe the "time-dilation" effect. Because of the inadequacy of the Galilean transformation to describe time dilation, Einstein used the Lorentz transformation in his special theory of relativity. The Lorentz transformation describes how two observers' varying measurements of space and time can be calculated (reconciled), even when those frames of measurement approach the speed of light. According to the Lorentz transformation, the clock on a jet plane would run slower than the clock at the train station as the plane flies overhead. This time dilation effect is science fact. It defies our normal experiences because we do not move at velocities that approach the speed of light. Therefore, we have no basis for an intuitive understanding of these phenomena.

Given the above information, let's circle back to our question: what is time? No one has defined it exactly. Most scientists, including Einstein, considered time (t) the numerical orders of physical events (change). The forth coordinate ($X_4 = ict$) is considered to be a spatial coordinate, on equal footing with X_1, X_2, and X_3 (the typical coordinates of three-dimensional space). With the advent of Einstein's special theory of relativity, time and space coalesce into space-time. Time is no longer an absolute. The Galilean transformation is set aside in favor of the Lorentz transformation, which is still in use today. At this point, all would seem as it should, regarding our perception of time.

However, in our thought experiment of an atom hypothetically frozen at absolute zero (after we suspended the laws of physics), the atom had no change. This poses a problem for the current definition of time as the numerical order of physical events. Since the atom is at absolute zero in our thought experiment, we have no change or order of physical events to measure. Yet, it continues to exist, and observers continue to evaluate its existence parallel to their own existence. Where does all this leave us regarding the nature of time?

We are at a point where we need to use our imagination and investigate a different approach to understand the nature of time. This is going to be speculative. After consideration, I suggest understanding the nature of time requires we investigate the kinetic energy associated with moving in four dimensions. The kinetic energy refers to an object's energy due to its movement. For example, you may be able to bounce a rubber ball softly against a window without breaking it. However, if you throw the ball at the window, it may break the glass. When thrown hard at the wall, the ball has more kinetic energy due to its higher velocity. The velocity described in this example relates to the ball's movement in three-dimensional space (X_1, X_2, and X_3). Even when the ball is at rest in three-dimensional space (like our atom at absolute zero), it is it still moving in the fourth dimension, X_4. This leads to an interesting question. If it is moving in the fourth dimension, X_4, what is the kinetic energy associated with that movement?

To calculate the kinetic energy associated with movement in the fourth dimension, $X_{4,}$ we use relativistic mechanics, from Einstein's special theory of relativity and the mathematical discipline of calculus.

Intuitively, it seems appropriate to use relativistic mechanics, since the special theory of relativity makes extensive use of Minkowski space and the X_4 coordinate, as described above. It provides the most accurate methodology to calculate the kinetic energy of an object, which is the energy associated with an object's movement.

If we use the result derived from the relativistic kinetic energy (Appendix I), the equation becomes:

$$KE_{X4} = -.3mc^2$$

Where KE_{X4} is the energy associated with an object's movement in time, m is rest mass of an object, and c is the speed of light in a vacuum.

For purposes of reference, I have termed this equation, $KE_{X4} = -.3mc^2$, the "Existence Equation Conjecture." Before you get a headache from this mind-blowing equation, let me attempt to calm your concerns. I will interpret the equation in significant detail below. In order to understand its derivation, you will need an aptitude with algebra, calculus, and an understanding of Einstein's equation for kinetic energy, which he derived from the special theory of relativity. (As a side note, Einstein developed the equations in the special theory of relativity, including $E = mc^2$, using algebra, calculus, empirical data, and thought experiments.) With the tools of algebra, calculus, and Einstein's equation for kinetic energy, along with the assumption that the object is at rest, the derivation is relatively straightforward. Without understanding these mathematical disciplines, the derivation becomes confusing. Don't overly concern yourself with the mathematics. However, it is more important to understand the fundamental concepts the equation is implying, rather than being able to derive or prove it. In Appendix II, I will provide data that demonstrates the Existence Equation Conjecture is able to accurately predict the life of an elementary particle (a muon), when it is accelerated close to the speed of the light. The prediction aligning with the experimental evidence is important, and provides evidence that the equation has a high likelihood of being correct. However, the word "conjecture," is part of the name for a reason. The scientific community needs to weigh in on its validity. Next, we will examine the features and implications of the equation.

Although, this equation is dimensionally correct (expressible in units of energy), which is a crucial test in physics, the equation is highly unusual from two standpoints. First, the kinetic energy is negative. The kinetic energy is always a positive value for real masses moving in three-dimensional space. However, as discussed previously, it can be negative for virtual particles. Second, the amount of negative kinetic energy suggested by the equation is enormous, approximately equal to a nuclear bomb, but negative in value.

How can we interpret this? Consider these concepts. Some aspects of time dilation (discussed in depth below) suggest adding energy in the form of positive kinetic energy or gravitational energy can extend the life of a subatomic particle. (Note: Gravitational energy can be added to any mass by placing it close to anther larger mass. For example, a clock close to the surface of the earth absorbs gravitational energy from the Earth.) Does this addition of energy feed the negative kinetic energy of the particle as it moves in the fourth dimension, thus extending the particle's life? Although Einstein never claimed that the fourth dimension was time, it appears related to time. For this reason, further exploration is appropriate.

Earlier in this chapter, we discussed the concept of space-time, and briefly touched on the effects of time dilation. To further our understanding of the nature of time, we will need to understand time dilation. The theory of time dilation has been around for about a century. It results from Einstein's special theory of relativity (circa 1905), and his general theory of relativity (circa 1916). What is time dilation? It is the difference of elapsed time between two events as measured by different observers, when the observers are moving relative to each other or the events. Time dilation also occurs when the observers are in stronger or weaker gravitational fields, relative to each other.

Here are two examples to illustrate dilation.

1) Picture yourself in a spaceship moving away from another observer who is at rest. When you look at your clock, it appears to be running normally. When the observer at rest looks at your clock, it appears to be running slower than his clock at rest. If the speed of your spaceship approaches the speed of light, the difference between the clocks is significantly exaggerated. The clock on the spaceship, from the viewpoint of the observer at rest, appears to have almost stopped.

2) Let's explore gravitational time dilation. When Einstein developed his general theory of relativity (circa 1916), he developed the theory of gravitational time dilation. Picture yourself in a spaceship near the sun, and another observer on a spaceship near the earth. To simplify things, assume you are both at rest relative to each other, and that each of you has a telescope capable of seeing the clock on the other's spaceship. The clock on the spaceship nearer the sun (in a much greater gravitational field) will move slower than the clock on the spaceship near the earth (in a lesser gravitational field). The observer near the sun sees his clock moving normally, but sees the observer's clock near the earth moving faster. The observer near the earth sees his clock moving normally, but sees the clock on the spaceship near the sun moving slowly.

Sounds like science fiction, but it is not. Time dilation is an experimentally verified fact. We are dealing with science fact, not science fiction. It is independent of the technical aspects of clocks, and not related to the speed of the signals, which is typically the speed of light. Science believes it is a fundamental of reality. Unfortunately, numerous academic derivations of the time-dilation equations are based on the relative motion of the observers, and the finite speed of the signals. This is misleading, and does not fit the experimental data. It leads to the incorrect conclusion that the time dilation effect is reciprocal (each observer observes the other observer's clock moving slower). This is how I first learned time dilation in 1966. It made a lot of sense, but not much experimental evidence was available to set the record straight. Unfortunately, although the derivation is simple, and provides a commonsense way to understand time dilation, it is misleading.

Evidence says the effects of kinetic energy and gravitation are additive. For example, a clock moving at near the speed of light, V_1, in a strong gravitational field, FG_1, will run slower than one moving at V_1 without a gravitational field or one at rest in a gravitational field FG_1.

Readers that wish to review the experimental evidence that proves time dilation is real may consult Appendix III ("Time Dilation Experimental Evidence").

The empirical evidence demonstrates that time dilates, slows down, by adding kinetic energy or gravitational energy. Does this help us interpret the Existence Equation Conjecture we have developed to determine the kinetic energy of a mass as it moves in the fourth dimension? Perhaps, However, the interpretation is going to be speculative and imaginative. This is okay. Recall that Einstein considered imagination crucial to scientific progress. With this caveat, here is one interpretation: Movement in the fourth dimension is associated with existence, and requires negative kinetic energy. This is similar to the positive kinetic energy required for movement in the typical three spatial dimensions. The difference is movement in the three spatial coordinates requires positive kinetic energy, while movement in the fourth spatial coordinate (existence) requires enormous negative energy, as suggested by the Existence Equation Conjecture ($KE_{X4} = -.3mc^2$). When we add kinetic energy or gravitational energy to a particle, we feed the negative energy that it requires to exist with the positive kinetic energy or gravitational energy. The negative kinetic energy of existence may be syphoning a portion of its energy from the particle. For a relatively small unstable particle at rest, such as a muon, we describe this existence as the expected life of the particle. If we add kinetic or gravitational energy to the particle, the negative kinetic energy of existence consumes less of the particle. Therefore, it increases its life.

This theoretical interpretation appears to fit the evidence presented regarding time dilation. For example, a muon at rest has an expected life in the order of 10^{-6} seconds. However, when muons naturally form via comic-ray collisions with our atmosphere, the resultant muon travels at speeds close to the speed of light before it reaches the ground. Therefore, its kinetic energy (KE) becomes extremely high. According to this interpretation, this high kinetic energy is providing the muon a portion of the energy required to exist. Therefore, it is increasing its expected life. This is consistent with the Rossi and Hall experiment performed in 1941, and the 1963 Frisch and Smith confirmation of their findings. (Descriptions of both experiments are in Appendix III.)

What does this suggest about the nature of time? According to our interpretation above, time is a measure of existence—and existence requires negative kinetic energy. Therefore, a relationship between time and energy exists. Is this too far out to be believable? I agree it stretches credibility to the limit. However, although the special theory of relativity has provided excellent equations to calculate time dilation, insight into the physical process behind time dilation remains elusive. College professors may teach special relativity in physic classes using concepts that involve the finite speed of light and relative motion of observers, but this does not explain the physical process. This teaching methodology would argue that time dilation is an optical illusion created by the motion of an observer relative to the light source. If this were strictly true, it would be reciprocal, and rule out the twin paradox. Based on the experimental evidence, the twin paradox is real. It appears that energy plays a role in time dilation. Specifically, adding kinetic energy and/or gravitational energy to a particle causes time dilation. The Existence Equation Conjecture may provide a framework to understand the actual physics behind time dilation.

It's still a dilemma. Where does the energy come from if existence requires negative energy? A simple examination of the Existence Equation Conjecture suggests the energy required for even a small mass, like an apple, to exist, would be equivalent to a nuclear bomb. In addition to that dilemma, the Existence Equation Conjecture suggests the energy expended to exist is negative. Theoretical physics has postulated the existence of negative energy. In 1930, the Dirac sea was postulated to reconcile the negative-energy quantum states, as predicted by Dirac in his mathematical modeling of electrons. However, science has not found a way to create negative energy. Currently, scientists are exploring the Casimir-Polder effect as a potential generator for negative energy. This may eventually yield fruitful results. This leaves us with a significant unanswered question. Where does the enormous negative energy required for existence come from?

There are relatively few candidates. In fact, after much thought and research, two emerge.

The gravitational fields of the universe. Gravitation's reach is infinite. Everything in the universe pulls on everything else. However, to date, no experimental evidence supports that the energy for existence is being syphoned from the universe's gravitational fields. Although, we know gravitational energy will cause time dilation, and extend a particle's existence, we do not observe any reduction in gravitational fields between objects. In fact, the force of gravity, as measured on Earth, remains unchanged over centuries of measurement. It is theoretically possible that dark matter blocks any gravitational energy syphoning within a galaxy, but its absence between galaxies allows the gravitational fields between galaxies to be syphoned, and thus causes the distance between galaxies to increase. This squares with observation, but this hypothesis has a serious problem: most distant galaxies are moving away from us at speeds that exceed the speed of light. If this is due to weakening gravitational fields between galaxies, it suggests the galaxies themselves are moving faster than the speed of light, and that violates the special theory of relativity, making it unlikely. Based on the above reasoning, until new data is available to the contrary, syphoning energy from gravitational fields between galaxies does not appear to be a viable candidate.

Dark energy. Is it possible that the universe's expansion is required to sustain existence? We know that the accelerated expansion of the universe is real. We attribute its expansion to an unknown cause: dark energy. We believe that dark energy is a kind of negative energy (acting repulsively) that causes space to expand. Consider this intriguing observation. Dark energy is a source of negative energy causing space to expand, and existence requires negative energy. Are the two related? This seems to suggest they are. In addition, the expansion of space, causing the most distance galaxies to move away faster than the speed of light, does not violate special relativity. It is not the galaxies moving, but rather the expansion of space. No law of physics prohibits the expansion of space from occurring faster than the speed of light. Thus, the expansion of space does not violate any physical laws. Dark energy may be a viable candidate as an energy source for existence.

There may be a relationship between entropy (a measure of disorder) and the Existence Equation Conjecture. What is the rationale behind this statement? First, time is a measure of change. Second, any change increases entropy in the universe. Thus, the universe's disorderliness is increasing with time. If we argue the entropy of the universe was at a minimum the instant prior to the Big Bang—since it represented an infinitely dense-energy point prior to change—then all change from the Big Bang on, served to increase entropy. Even though highly ordered planets and solar systems formed, the net entropy of the universe increased. Thus, any change, typically associated with time, is associated with increasing entropy. This implies that the Existence Equation Conjecture may have a connection to entropy.

One last question requires examination. Why is the expansion accelerating? This may be due to the increasing complexity of the universe. With the near infinite number of objects in the universe existing and changing, the amount of change over time has increased, and even accelerated.

What does all of the above say about the nature of time? If we are on the right track, it says describing the nature of time requires six crucial elements, all of which are simultaneously true.

1) Time is change. (This is true, even though it was not true in our "thought experiment" of an isolated atom at absolute zero. As mentioned above, it is not possible for any object to reach absolute zero. The purpose of the thought experiment was to illustrate the concept of "existence" separate from "change.")

2) Time is a measure of energy, since change requires energy.

3) Time is a measure of existence. (The isolated atom, at absolute zero, enables us to envision existence separate from change.)

4) Movement in time (or existence) requires negative energy.

5) The energy to fuel time (existence) is enormous. It may be responsible for the life times associated with unstable elementary particles, essentially consuming them, in part, to satisfy the Existence

Equation Conjecture. It may be drawing energy from the universe (dark energy). If correct, it provides insight into the nature of dark energy. Essentially the negative energy we call dark energy is required to fuel existence.

6) Lastly, the enormousness changes in entropy, creating chaos in the universe, may be the price we pay for time. Since entropy increases with change, and time is a measure of change, there appears to be a time-entropy relationship. In addition, entropy proceeds in one direction. It always increases when change occurs. Time proceeds in one direction, from the present to the future. Philosophers term this the arrow of time. The directional alignment, and the physical processes of time, suggests a relationship between time and entropy.

This theory of time is speculative, but fits the empirical observations of time. A lot of the speculation rests on the validity of the Existence Equation Conjecture. Is it valid? As shown in Appendix II, it is entirely consistent with data from a high-energy particle-accelerator experiment involving muons moving near the speed of light. The experimental results agree closely with predictions of the Existence Equation Conjecture (within 2%). This data point is consistent with the hypothesis that adding kinetic energy can fuel the energy required for existence. The implications are enormous, and require serious scientific scrutiny. I published the Existence

Equation Conjecture in this book to disseminate information, and enable the scientific scrutiny.

The Existence Equation Conjecture represents a milestone. If further evaluation continues to confirm the validity of the Existence Equation Conjecture, we have a new insight into the nature of time. Existence (movement in time) requires enormous negative energy. The Existence Equation Conjecture, itself, provides insight into the physical processes underpinning time dilation. It answers the question why a subatomic particle's life increases with the addition of kinetic or gravitational energy. It offers a solution path to a mystery that has baffled science since 1998, namely the cause of the accelerated expansion of the universe. Lastly, it may contain one of the keys to time travel.

Now that we have explored the nature of time, I would like invite you to take a trip with me. It will be an exciting adventure. Before you answer, let me warn you. Although no law of physics blocks our way, the road is unknown. Navigation will be difficult and maybe even perilous. In fact, we may not return. If we do return, we do not know what we will find when we return. The world as we know it may have irrevocably changed. It is not a short trek over land or sea, nor will we journey into space. Perhaps you have guessed my proposed adventure. I am inviting you to travel in time with me. "Impossible." you might say. I am not sure. In fact, substantial scientific evidence says that it is possible. Perhaps, we are a little ahead of our time. We will consider the possibility of time travel in the next chapter.

"Once confined to fantasy and science fiction, time travel is now simply an engineering problem."

Michio Kaku, Wired Magazine, Aug. 2003

CHAPTER 13

Is Time Travel Possible?

Few topics in science capture the imagination like time travel. Science fiction, like H. G. Wells' classic novel, *The Time Machine,* published in 1895, and science fact, like time dilation, continues to fuel interest in time travel. In the last chapter, we completed our exploration of the nature of time. This positions us to address this chapter's main question. Is time travel possible?

Of course, time travel is possible. We are already doing it. At this point, you know I can be a bit flippant. However, as usual, my answer has a kernel of truth. We are traveling in time. We continually travel from the present to the future. This is what philosophers refer to as the arrow of time. In our everyday experience, it moves in one direction, from the present to the future. I think, though, on a more serious note, what people want to know is can we travel back in time—or to a future date in time.

In theory, it is possible. Indeed, numerous solutions to Einstein's general relativity equations predict time travel is possible. In general, no law of physics prohibits time travel. We will begin by considering several methods science proposes in order to travel in time.

Faster-than-light (FTL) time travel to the future

Using faster than light or near the speed of light, time travel appears to offer methodologies grounded in science fact. Consider two examples:

1) Assume you build a spaceship capable of traveling near the speed of light. With such a spaceship, you literally can travel into the future. This may sound like science fiction, but it is widely accepted as scientific fact. Particle accelerators confirm it. We discussed it when we discussed time dilation and the twin paradox. All you need is the spaceship, and an enormous amount of energy to accelerate it near the speed of light. However, this is an enormous problem. From Einstein's special theory of relativity, we know that as we begin to accelerate a mass close to the speed of light, it becomes more massive, and approaches infinity. Thus, to accelerate it close to the speed of light, we need an energy source that approaches infinity. Perhaps we would have to learn

how to harness the energy of a star, or routinely create matter-antimatter annihilations to create energy. Today's science is nowhere near that level of sophistication.

2) Assume you can move information (like a signal) faster than light. Theoretically, if we could send a signal from point A to point B faster than the speed of light, it would represent a form of time travel. However, a significant paradox occurs. Here is an example.

   An observer A in an inertial frame A sends a signal to an observer B in an inertial frame B. When B receives the signal, B replies and sends a signal back to A faster than the speed of light. Observer A receives the reply before sending the first signal.

In 1907, Albert Einstein described this paradox in a thought experiment to demonstrate that faster-than-light communications can violate causality (the effect occurs before the cause). Albert Einstein and Arnold Sommerfeld in 1910 described a thought experiment using a faster-than-light telegraph to send a signal back in time. In 1910, no faster-than-light signal communication device existed. It still does not exist, but the possibility of its development is increasing. We know from quantum physics that certain quantum effects "transmit" instantaneously and, therefore, faster than the speed of light in empty space. One example of this is the quantum states of two "entangled" particles (particles that have physically interacted, and later separated). In quantum physics, the quantum state is the set of mathematical variables that fully describes the physical aspects of a particle at the atomic level. When two particles interact with each other, they appear to form an invisible bond between them. When this happens, they become "entangled." If we take one of the particles, and separate it from the other, they remain entangled (invisibly connected). If we change the atomic state of one of the entangled particles, the other particle instantaneously changes its state to maintain quantum-state harmony with the other entangled particle. Significant experimental evidence indicates that separated entangled particles can instantaneously transmit information to each other over distances that suggest the information exchange exceeds the speed of light. Initially, scientists criticized the theory of particle entanglement. After its experimental verification, science recognizes entanglement as a valid, fundamental feature of quantum mechanics. Today the focus of the research has changed to utilize its properties as a resource for communication and computation.

Using wormholes

Scientists have proposed using "wormholes" as a time machine. A wormhole is a theoretical entity in which space-time curvature connects two distant locations (or times). Although we do not have any concrete evidence that wormholes exist, we can infer their existence from Einstein's general theory of relativity. However, we need more than a wormhole. We need a traversable wormhole. A traversable wormhole is exactly what the name implies. We can move through or send information through it.

If you would like to visualize what a wormhole does, imagine having a piece of paper whose two-dimensional surface represents four-dimensional space-time. Imagine folding the paper so that two points on the surface are connected. I understand that this is a highly simplified representation. In reality, we cannot visualize an actual wormhole. It might even exist in more than four dimensions.

How do we create a traversable wormhole? No one knows, but most scientists believe it would require enormous negative energy. This is interesting, since the Existence Equation Conjecture, derived previously, implies moving in time requires negative energy. A number of scientists believe the creation of negative energy is possible, based on the study of virtual particles and the Casimir effect.

Assuming we learn how to create a traversable wormhole, how would we use it to travel in time? The traversable wormhole theoretically connects two points in space-time, which implies we could use it to travel in time, as well as space. However, according to the theory of general relativity, it would not be possible to go back in time prior to the creation of the traversable wormhole. This is how physicists like Stephen Hawking explain why we do not see visitors from the future. The reason: the traversable wormhole does not exist yet.

Stephen Hawking did a fascinating time-traveler experiment in his popular TV series, "Into the Universe with

Stephen Hawking." He held a reception for time travelers from the future. He sent the invitations out after the reception had already occurred. His hope was that someone in the far-distant future would come across the invitation, and travel back in time to attend the reception. In the TV series, you see the reception room and Stephen Hawking, but no time travelers. He was disappointed.

However, we have four possible explanations why no time travelers attended:

1) The invitation did not survive into the far-distant future, a future whose science enabled time travel to the past.
2) Time travel into the past is not possible in the future, regardless of how far into the future the invitations survive.
3) The human race does not exist in the distant future, destroyed by our own hand, or a cosmic calamity.
4) Time travelers showed up at the party, but it was in another universe (an alternate reality suggested by the "Many-Worlds of Quantum Mechanics" theory). Perhaps in that reality, the TV series broadcasts a reception room filled with time travelers.

Although, we are discussing time travel, it is essential to note that wormholes imply connections between different points in space. This means that they may provide a faster-than-light connection between two planets, for example. Although faster-than-light travel is not possible, the wormhole may represent a shortcut. Travel inside the wormhole may remain below the speed of light, but be faster than the time it would take light to traverse the same two points outside the wormhole. Think of this simple picture.

> You are on one side of the mountain. If you want to travel to the other side of the mountain by traversing its circumference, the journey will take longer than using a tunnel that connects to the other side of the mountain. The speed you travel is the same, but the tunnel allows a shortcut, and it appears that you traveled faster.

Will we ever be able to create traversable wormholes? Theoretically, it appears possible. Experiments are being conducted, as I write, using the Large Hadron Collider to create small wormholes, small black holes, and dark matter. The next decade holds considerable promise to address these questions.

Using black holes

Scientists have postulated that we could use a black hole, which already has the energy we need, as a time machine. Unfortunately, the nearest black hole we know of is at the center of our galaxy. Doing the experiment is difficult. First, we have the enormous problem of getting to the black hole. The Milky Way galaxy is 100,000-120,000 light-years in diameter. Therefore, given the position of our planet in the galaxy, we are about 40-50 million light-years away from the center, where the black hole is located. Once we get there, we would have to deal with the crushing gravitational pull that would compress us to about the size of subatomic particle or smaller, or might even turn us into energy. None of the outcomes appears to satisfy our quest.

Numerous solutions to Einstein's general-relativity equations predict time travel is possible. Within the scientific community, various concepts describe possible ways to do it. So, why are we still stuck in the present? Although the solutions are theoretically possible, we do not have the science. Most scientists surmise that the energy requirements for time travel are enormous, and potentially exotic. Using today's science, we do not know how to harness the energy of a star in order to power a spaceship to near the speed of light, create a traversable wormhole, or go to and survive a black hole ride.

Time travel to the past appears to represent the greatest challenge. First, as opposed to time dilation, which is experimentally provable and gives us a "theoretical" methodology to travel to the future, we have no conclusive experimental evidence that suggests time travel to the past is possible. We have solutions to Einstein's general relativity equations that suggest it is possible, but the experimental evidence to date is questionable. The crucial issue regarding time travel to the past is that it suggests we can violate causality (cause and effect). This means that the effect would come before the cause, "reversed causality." To date, a number of controversial experiments have been performed to demonstrate reversed causality is possible. Important word here is "controversial." Although numerous experiments have been done to demonstrate reversed causality, most have been dismissed as measurement errors or explained while using known physics, like quantum mechanics.

However, one physicist, Ronald Lawrence Mallett, is making time travel a cornerstone of his research. Although his work is controversial, it has garnered worldwide attention. It is ongoing research and hotly debated. Since this research is topical, and the debate continues as I write, I felt it makes an instructive case study.

Ronald Lawrence Mallett received his Ph.D. in 1973 from Penn State University. After obtaining a position at the University of Connecticut in 1975 to teach physics as an assistant professor, he progressed to become a full professor in 1980. In 2006, Dr. Mallet coauthored a book with New York Times best-selling author Bruce B. Henderson, *Time Traveler: A Scientist's Personal Mission to Make Time Travel a Reality.* In 2007, Dr. Mallett's life story of pursuing a time machine was featured on a weekly radio program, This American Life, episode 324, "My Brilliant Plan." In 2008, the movie rights to Dr. Mallett's book was purchased by Spike Lee, an American film director, producer, writer, actor, and winner of an Emmy award and nominated for two Academy Awards.

Dr. Mallet's innovative approach to causality violation involves passing a neutron through a twisted laser circle in a light cylinder. Dr. Mallet claims to have found new solutions of the Einstein field equations for the exterior and interior gravitational fields of the light cylinder—and that indicates closed timelike curves will result. Closed timelike curves suggest the possibility of time travel into the past. (Closed timelike curves are how physicists typically refer to time travel.)

Physicists Ken Olum and Allen Everett published an article (Olum, Ken D.; Allen Everett (2005). Their paper in Foundations of Physics Letters 18 (4): 379–385), *Can a circulating light beam produce a time machine?*, calls into question the scientific foundation of Dr. Mallet's experimental approach, and argues the amount of energy needed to twist space-time would be enormous. Lastly, they cited a theorem by Stephen Hawking, in his 1992 paper on the "Chronology Protection Conjecture," which they used to call into question Dr. Mallet's scientific basis for the experiment.

Dr. Mallet has answered some, but not all the objections raised by the Olum Everett paper. The 1992 paper, "Chronology Protection Conjecture," published by Stephen Hawking, has itself been highly scrutinized, with scientists taking pro and con positions. Several physicists dispute Hawking's proof, which essentially demonstrates wormholes would be unstable. Sergei Krasnikov, a Russian physicist quoted in Michio Kaku's book Parallel Worlds (Anchor Books 2006), argued, "there is not a grain of evidence to suggest that the time machine (wormhole) must be unstable." Li-Xin Li, a Princeton physicist quoted in Dr. Kaku's Parallel Worlds book, published an antichronology protection conjecture, stating, "There is no law of physics preventing the appearance of closed timelike curves." In 1998, this remarkable backlash caused Hawkins to acknowledge that portions of the criticism are valid.

Where does this leave us with Dr. Mallet's research? The question remains open. Dr. Mallet continues his time travel research, surrounded by supporters and critics. Obviously, Dr. Mallet understands that his work is going to come under the scientific community's magnifying glass. Time travel experiments themselves are on "fringe" in the current scientific environment. This is not deterring Dr. Mallet. He is courageously moving forward.

The title of this chapter is a deceptively simple question. Is time travel possible? The answer is definitive. Yes! It is theoretically possible. No scientific theory to date rules out time travel. Indeed, the numerous solutions to Einstein's general theory of relativity regarding time travel make it seem as if we should already be doing it. Have you ever noticed that today's science fiction becomes tomorrow's science fact? I have lived long enough to see it happen. For example, the communicators on Star Trek are feeble compared to today's smart phones. Did you ever see Captain Kirk get an instant message or watch a movie on his communicator? Schoolchildren do text messaging and watch movies on their smart phones every day. Although the problems with time travel appear monumental and unsolvable by today's science, the problems may be straightforward and solvable in 100 years. Look at the advances made in science in the last century.

However, even if time travel is possible, what about time-travel paradoxes? If we travel in time, will we create time-travel paradoxes? Will the mere footprint of a time traveler alter our reality? Do time-travel paradoxes render time travel impossible? We will discuss this in the next chapter ("Time-Travel Paradoxes").

"I can not travel into my past, without consent of the future me."

Toba Beta, My Ancestor Was an Ancient Astronaut

CHAPTER 14

Time-Travel Paradoxes

What is a time-travel paradox? They are thought experiments that demonstrate time travel violates causality (cause and effect). We have no experimental evidence of time-travel paradoxes because we have been unable to build a time machine that allows us to travel to the future or to the past. However, as Einstein demonstrated throughout his career, thought experiments can yield deep insight into science. In fact, on one occasion when Einstein was asked the location of his laboratory, he smiled, took out his fountain pen, and said, "Here!"

We'll start with one of the most popular time-travel paradoxes, involving time travel to the past. It is the most-sited time-travel paradox, namely the "grandfather paradox." It goes something like this: You travel back in time, and prevent your grandfather from meeting your grandmother. Thus, they never meet, which means you are never born. Numerous time-travel paradoxes are similar to this. For example:

> Imagine you construct a wormhole, and see yourself through the wormhole prior to constructing the wormhole. Imagine that you shoot yourself through the wormhole before you (in the past) construct the wormhole. What happens to the wormhole? What happens to you? These two time-travel paradoxes have to do with time travel to the past.

Next, consider time-travel paradoxes caused by time travel to the future:

> Imagine, you travel to the future, and bring the cure for cancer back to your own time. Assume that the discovery to cure cancer took one hundred years from the time you left for the future. What affect does this have on the future? Millions of people, that might have died, will survive cancer. These people's lives have every possibility of altering the future you visited. If you time travel to the future again, what will it look like? Will there even be a future?

Let's consider another future time paradox:

> Assume your destiny is to marry and have children

> in your own time. However, suppose you time travel to the future before you marry and have children. Suppose you never return to your own time to fulfill your destiny. Have you changed the past? Do the history books instantly change?

This boils down to one crucial question. Do time-travel paradoxes make time travel impossible? It appears they do not. Numerous theories, themselves thought experiments, suggest resolutions to time-travel paradoxes. The theories appear to fall into four categories.

Category I: Multiverse Hypothesis

Time-travel paradoxes occur, but lead to alternate realties. The best-known theory in this class is Everett's many-worlds interpretation (MWI) of quantum mechanics. The majority of cosmologists and quantum-mechanics physicists believe this theory, discussed in Chapter 4. This theory states that all events lead to mutually exclusive histories. In effect, time branches off to accommodate all realities. This is one of the arguments for parallel worlds. One question haunts this theory. Would the time traveler be able to travel back to his own time, or would he be required to travel back to a different history, essentially a new parallel world? David Deutsch, British physicist at the University of Oxford, argues the time traveler must stay in the new parallel world. Stephen Hawking disagrees and argues that each time traveler should experience a single self-consistent history. Thus, the time traveler returns to his or her own time. These two highly notable physicists strongly disagree. What do you think?

Category II: Timeline-Protection Hypothesis

Time paradoxes are impossible. This theory suggests that if you go back to the past, and attempt to prevent your grandfather from meeting your grandmother, you fail every time. In other words, creating a time paradox is impossible. Somehow, your grandfather always meets your grandmother, regardless of any of your attempts. If you attempt to shoot yourself though a wormhole, the gun jams or something else happens that causes you to fail to create the paradox. Perhaps you even reconsider, and are unable to shoot yourself. In other words, you are able to travel in time, but unable to alter time. The Novikov self-consistency principle, suggested by Russian physicist Igor Dmitriyevich Novikov in the mid-1980s, fits within this category of theories. According to Novikov's principle, anything a time traveler does remains consistent with history. In other words, the time traveler is unable to change history. The self-healing hypothesis seems to fit under this class of theories. This theory states that whatever a time traveler does to alter the present, by traveling back in history, sets off another set of events to cause the present to remain unchanged. In effect, time heals itself.

Category III: Timeline-Corruption Hypothesis

This theory suggest that time paradoxes must occur, and in fact, are unavoidable. Any time travel creates minute effects that inevitably alter the timeline. Thus, if you inadvertently step on a small bug in the past, it changes the future. Popular science fiction literature calls this the "butterfly effect," namely that the flutter of a butterfly's wings in Africa can cause a hurricane in North America. Under this theory, anything you do will have a consequence. It may be small and benign. Alternatively, it may be large and disastrous. The destruction-resolution hypothesis fits in this class of theories. It holds that anything a time traveler does to cause a paradox, destroys the timeline (and perhaps the universe). Obviously, this is the "large and disastrous" outcome mentioned above.

Category IV: Choice Timeline Hypothesis

The choice timeline hypothesis holds that if you choose to travel in time, it is predestined, and history instantly changes. This implies you can time travel to the future, and leave an item there that you will need, sometime in the future. It will be there for you when the future becomes the present. To illustrate this, assume you are in New York City, and someone is about to assault you. You have no escape or means of protection. According to the choice timeline hypothesis, you use your time machine to travel to the future. You hide a gun near the place where the assault is about to occur. When the assault occurs, you simply retrieve the hidden gun, and scare off the attacker. Arguably, scientists hold the choice timeline hypothesis is most consistent with our understanding of the time dimension.

Of the numerous other time-paradox hypotheses, most fall under one of the above categories, or are not as popular as the above. I left them out in the interest of clarity and brevity. The four categories above give us a reasonable framework to understand time-travel paradoxes, and the current thinking regarding their impact on the timeline. The interesting aspect is that most of the scientific community does not think time paradoxes inhibit time travel. The scientific consensus appears to be that time paradoxes may or may not occur, but they do not exclude the possibility of time travel. In fact, Kip Thorne, an American theoretical physicist and Professor of Theoretical Physics at the California Institute of Technology until 2009, argues that time paradoxes are imprecise thought experiments, which can be resolved by numerous consistent solutions.

Are you ready for another mystery? As you read these words, the light reflects off the paper of the book or emits from the computer screen at an incredibly fast velocity. Why does light travel incredibly fast? That is the subject of our next mystery: "The Mysterious Nature of Light."

"All the fifty years of conscious brooding have brought me no closer to answer the question, 'What are light quanta?' Of course today every rascal thinks he knows the answer, but he is deluding himself."

Albert Einstein, (1951). Quoted in Raymond W. Lam, Seasonal Affective Disorder and Beyond

CHAPTER 15

The Mysterious Nature of Light

To almost everyone, there is nothing mysterious about light. In fact, the opposite is true. When we are in the dark and mystery abounds, the first thing we do is turn on the lights. So, why is "The Mysterious Nature of Light" the title of this chapter?

The first thing that makes light mysterious is that it can exhibit both the properties of a wave and a particle. For all of the Nineteenth Century, and for the early part of the Twentieth Century, most scientists considered light "a wave," and most of the experimental data supported that "theory." However, classical physics could not explain black-body radiation (the emission of light due to an object's heat). A light bulb is a perfect example of black-body radiation. The wave theory of light failed to describe the energy (frequency) of light emitted from a black body. The energy of light is directly proportional to its frequency. To understand the concept of frequency, consider the number of ocean waves that reach the shore in a given length of time. The number of ocean waves than reach the shore, divided by the length of time you measure them, is their frequency. If we consider the wave nature of light, the higher the frequency, the higher the energy.

In 1900, Max Planck hypothesized that the energy (frequency) of light emitted by the black body, depended on the temperature of the black body. When the black body was heated to a given temperature, it emitted a "quantum" of light (light with a specific frequency). This was the beginning of Quantum Mechanics. Max Planck had intentionally proposed a quantum theory to deal with black-body radiation. To Planck's dismay, this implied that light was a particle (the quantum of light later became known as the photon in 1925). Planck rejected the particle theory of light, and dismissed his own theory as a limited approximation that did not represent the reality of light. At the time, most of the scientific community agreed with him.

If not for Albert Einstein, the wave theory of light would have prevailed. In 1905, Einstein used Max Planck's black-body model to solve a scientific problem known as the photoelectric effect. In 1905, the photoelectric effect was one of the great unsolved mysteries of science. First discovered in 1887 by Heinrich Hertz, the photoelectric effect referred to the phenomena that electrons are emitted from metals and non-metallic solids, as well as liquids or gases, when they absorb energy from light. The mystery was that the energy of the ejected electrons did not depend on the intensity of the light, but on its frequency. If a small amount of low-frequency light shines on a metal, the metal ejects a few low-energy electrons. If an intense beam of low-frequency light shines on the same metal, the metal ejects even more electrons. However, although there are more of them, they possess the same low energy. To get high-energy electrons, we need to shine high-frequency light on the metal. Einstein used Max Planck's black-body model of energy, and postulated that light, at a given frequency, could solely transfer energy to matter in integer (discrete number) multiples of energy. In other words, light transferred energy to matter in discrete packets of energy. The energy of the packet determines the energy of the electron that the metal emits. This revolutionary suggestion of quantized light solved the photoelectric mystery, and won Einstein the Nobel Prize in 1921. You may be surprised to learn that Albert Einstein won the Nobel Prize for his work on quantizing light—and not on his more famous theory of relativity.

The second property of light that makes it mysterious is its speed in a vacuum. The speed of light in a vacuum sets the speed limit in the universe. Nothing travels faster than light in a vacuum. In addition, this is a constant, independent of the speed of the source emitting the light. This means that the light source can be at rest or moving, and the speed of light will always be the same in a vacuum. This is counterintuitive. If you are in an open-top convertible car speeding down the highway, and your hat flies off, it begins to move at the same speed as the car. It typically will fall behind the car due to wind resistance that slows down its speed. If you are in the same car, and throw a ball ahead of the car, its velocity will be equal to the speed of the car, plus the velocity at which you throw it. For example, if you can throw a ball sixty miles per hour and the car is going sixty miles per hour, the velocity of the ball will be one hundred twenty miles per hour. This is faster than any major league pitcher can throw a fastball. Next, imagine you are in the same car and have a flashlight. Whether the car is speeding down the highway or parked, the speed of light from the flashlight remains constant (if we pretend the car is in a vacuum). The most elegant theory of all time, Einstein's special theory of relativity, uses this property of light as a fundamental pillar in its formulation.

- Why does light have a wave-particle duality?
- Why is the speed of light in a vacuum the upper limit of anything we observe in the universe?
- Why is the speed of light a constant independent of the movement of the source emitting the light?

No one knows. We learned an enormous amount about light in the last hundred years. We know it is composed of photons (packets of energy) that have no mass, and when emitted instantaneously, they travel at exactly 299,792,458 meters per second—about 186,000 miles per second. This means they do not accelerate to that speed. They instantaneously exist at that speed. We know the speed of light is a constant independent of the velocity of the source that emits the light. Lastly, we know photons can exhibit the properties of a wave and a particle. The one thing we do not know is "why."

It is time to move on to our next mystery. In a way, you can say it is about money, but not the kind of money you carry in your pocket. Perhaps, a better word would be to say it is about "currency." I am not talking about gold, silver, or diamonds, all of which people describe as a universal currency. Almost everyone universally considers these valuable. However, what do I mean by currency? Energy! The universal currency of the universe is energy. Advanced aliens, if they exist, may not understand anything about our monetary currencies, but they will likely understand everything we do to create and harness energy. These aliens will likely value energy, and closely guard the processes they use to create and harness energy. The critical importance of energy is relatively easy to illustrate. Consider these few questions. Can you run your car on gold? No, it requires gasoline, which the car's engine converts to energy. Can you eat silver or gold to live? No, you need food and water, which your body transforms into energy. If advanced aliens exist, do you think they traverse the universe looking for gold, silver, or diamonds? Although gold, silver, or diamonds may have industrial uses for them, I think their real quest would be energy. To advanced aliens, energy would have high value. Energy would make interstellar travel possible. Energy would allow them to mine any planet for whatever they chose. With energy, plus technology, they could do and get anything they wanted. This is why I believe energy is the universal currency of the universe. A logical question naturally arises. What is energy? That is the problem. We do not know. Energy is an enigma, and the subject of our next chapter.

We are quite ignorant of the condition of energy in bodies generally. We know how much gas goes in, and how much comes out, and know whether at entrance and exit it is in the form of heat or of work. That is all.

Peter Guthrie Tait,
Sketch of Thermodynamics (1877), 137.

CHAPTER 16

The Energy Enigma

We scientists talk about energy, and derive equations with energy mathematically expressed in the equation as though we understand energy. The fact is: we do not. It is an indirectly observed quantity. We infer its existence. For example, in physics, we define energy as the ability of a physical system to do work on another physical system. Physics is one context that uses and defines the word energy. However, the word energy has different meanings in different contexts. Even the average person throws the term energy around in phrases like, "I don't have any energy today," generally inferring a lack of vigor, force, potency, zeal, push, and the like. The word energy finds its way into both the scientific community and our everyday communications, but the true essence of energy remains an enigma.

The concept of energy is an old concept. It comes from the ancient Greek word, "enérgeia," which translates "activity or operation." As previously stated, we do not know the exact essence of energy, but we know a great deal about the effects of energy. To approach a better understanding, consider these four fundamental properties of energy:

1. Energy is transferable from one system to another.

Transferring mass between systems results in a transfer of energy between systems. Mass and energy have been inseparably equated, since 1905, via Einstein's famous mass–energy equivalence equation, $E = mc^2$, where E is energy, m is mass, and $\boldsymbol{c}$ is the speed of light in a vacuum. This equation is widely held as a scientific fact. Experimental results over the last century strongly validate it. Typically, mass transfers between systems occur at the atomic level as atoms capture subatomic particles or bond to form products of different masses.

Non-matter transfer of energy is possible. For example, a system can transfer energy to another by thermal radiation (heat). The system that absorbs the thermal radiation experiences an increase in energy, typically measured by its temperature. This is how the radiators in a house raise the room temperature. Here is another example: If an object in motion strikes another object, a transfer of kinetic energy results. Consider billiard balls. When one ball strikes another, it imparts kinetic energy to the ball it strikes, causing it to move.

2. Energy may be stored in systems.

If you pick up a rock from the ground and hold it at shoulder height, you have stored energy between the rock and ground via the gravitation attraction created between the Earth and rock. You may consider this potential energy. When you open your hand, the rock will fall back to the ground. Why? The answer is straightforward. It required your energy to hold the rock in its new position at shoulder height. As soon as you, by opening your hand, released the energy that you were providing, it reduced to a lower energy state when the gravitational field pulled the rock back to the ground.

Any type of energy that is stored is "potential energy," and all types of potential energy appear as system mass. For example, a compressed metal spring will be slightly more massive than before it was compressed. When you compress the spring, you do work on the system. The work on the system is energy, and that energy is stored in the compressed spring as potential energy. Because of this stored potential energy, the spring becomes more massive.

3. Energy is not only transferable–it is transformable from one form to another.

Our example regarding the rock falling back to the ground is an example of energy transformation. The potential energy was transformed to kinetic energy when you opened your hand and released the rock. This is what caused the rock to fall back to the ground. Here is an industrial example. Hydroelectric plants generate electricity by using water that flows over a falls due to gravity. In effect, they are transforming the falling water (gravitational energy) into another form of energy (electricity).

4. Energy is conserved.

This is arguably the most sacred law in physics. Simply stated: Energy cannot be created or destroyed in an isolated system. The word "isolated" implies the system does not allow other systems to interact with it. A thermos bottle is an example of an isolated system. It is preventing the ambient temperature from changing the temperature inside the thermos. For example, it keeps your coffee hot for a long time. Obviously, it is not a perfectly isolated system since eventually it will lose heat to the atmosphere, and your coffee will cool to the ambient temperature that surrounds the thermos bottle. For example, in your house, the coffee in a cup will cool to room temperature.

In summary, energy may be transferred, stored, and transformed, but it cannot be created or destroyed in an isolated system. This means the total energy of an isolated system does not change.

Next, we will consider energy in different contexts. Unfortunately, since we do not know the true essence of energy, we need to describe it via the effects we observe in the context that we observe them. Here are two contexts:

1) Cosmology and Astronomy

Stars, nova, supernova, quasar, and gamma-ray bursts are the highest-output mass into energy transformations in the universe. For example, a star is typically a large and massive celestial body, primarily composed of hydrogen. Due to its size, gravity at the star's core is immense. The immense gravity causes the hydrogen atoms to fuse together to form helium, which causes a nuclear reaction to occur. The nuclear reaction, in effect, transforms mass into energy. In the cosmos, mass-to-energy transformations are due to gravity, and follow Einstein famous equation, $E = mc^2$ (discussed previously). The gravity can result in nuclear fusion, as described in the above example. It can cause a dying star to collapse and form a black hole.

2) Chemistry

Energy is an attribute of the atomic or molecular structure of a substance. For example, an atom or molecule has mass. From Einstein's mass-energy equivalence equation, ($E = mc^2$), we know the mass equates to energy. In chemistry, an energy transformation is a chemical reaction. The chemical reaction typically results in a structural change of the substance, accompanied by a change in energy. For example, when two hydrogen atoms bond with one oxygen atom, to form a water molecule, energy emits in the form of light.

Other scientific contexts give meaning to the word energy. Two examples are biology and geology. Numerous forms of energy are accepted by the scientific community. The various forms include thermal energy, chemical energy, electric energy, radiant energy, nuclear energy, magnetic energy, elastic energy, sound energy, mechanical energy, luminous energy, and mass. I will not go into each form and context for the sake of brevity. My intent is to illustrate that the word energy in science must be understood within a specific context and form.

As mentioned above, we truly do not know the essence of energy; we infer its existence by its effects. The effects we measure often involve utilizing fundamental concepts of science, such as mass, distance, radiation, temperature, time, and electric charge. Adding to ambiguity, energy is often confused with power. Although we often equate "power" and "energy" in our everyday conversation, scientifically they are not the same. Strictly speaking, in science, power is the rate at which energy is transferred, used, or transformed. For example, a 100-watt light bulb transforms more electricity into light than a 60-watt light bulb. In this example, the electricity is the energy source. Its rate of use in the light bulb is power. It takes more power to run a 100-watt bulb than a 60-watt bulb. Your electric bill will verify this is true.

What is it about energy that makes it mysterious? Science does not understand the nature of energy. We have learned a great deal about energy in the last century. The word energy has found its way into numerous scientific contexts as well as into our everyday vernacular, but we do not know the fundamental essence of energy. We can infer it exists. Its existence and definition is context sensitive. We do not have any instrument to measure energy directly, independent of the context. Yet, in the last century, we have learned to harness energy in various forms. We use electrical energy to power numerous everyday items, such as computers and televisions. We have learned to unleash the energy of the atom in nuclear reactors to power, for example, cities and submarines. We have come a long way, but the fundamental essence of energy remains an enigma.

In the next chapter, we will discuss another aspect of energy that haunts the scientific community. Does all reality consist of discrete packets (quantums) of energy? Are mass, space, time, and energy composed of quantized energy? We can make a reasonably strong case that they are. It is counterintuitive because we do not experience reality that way. For example, when you pick up a rock, you do not directly experience the atoms that make up the rock. However, the rock is nothing more than the sum of all its atoms. If all reality is made of quantized energy, we live in a Quantum Universe. What exactly is a Quantum Universe? The answer is one page away, so read on.

"Reality is merely an illusion, albeit a very persistent one."

Albert Einstein (1879-1955)

CHAPTER 17

The Quantum Universe

The notion that all reality (mass, space, time, and energy) consists of discrete energy quantums is counterintuitive. For example, an electric current consists of individual electrons flowing in a wire. However, you do not notice your television flickering as the electrons move through the circuits. The light you read by consists of individual photons. Yet, your eyes do not sense individual photons reflected from the page. The point is that our senses perceive reality as a continuum, but this perception is an illusion. In the following, we will examine each element of reality one by one to understand its true nature.

Mass—the sum of all its atoms.

We will start with mass. Any mass is nothing more than the sum of all its atoms. The atoms themselves consist of subatomic particles like electrons, protons, and neutrons, which consist of even more elementary particles, like quarks. (Quarks are considered the most elementary particles. I will not describe the six different types of quarks in detail, since it will unnecessarily complicate this discussion.) The point is any mass reduces to atoms, which further reduces to subatomic particles. The atom is a symphony of these particles, embodying the fundamental forces (strong nuclear, weak nuclear, electromagnet, and gravity). Does all this consist of energy quantums? In the final analysis, it appears it does, including the fundamental forces themselves. How can this be true?

In the early part of the Twentieth Century, the theory of quantum mechanics was developed. It is able to predict and explain phenomena at the atomic and subatomic level, and generally views matter and energy as quantized (discrete particles or packets of energy). Quantum mechanics is one of modern science's most successful theories. At the macro level, which is our everyday world, any mass is conceivably reducible to atoms, subatomic particles, and fundamental forces.

Science holds that the fundamental forces (strong nuclear, weak nuclear, electromagnet, and gravity) mediate (interact) via particles. For example, the electromagnetic force mediates via photons. We have verified the particle for all the fundamental forces, except gravity. A number of theoretical physicists believe a particle is associated with gravity, namely the graviton. The graviton is a hypothetical elementary massless particle that theoretical physicists believe is responsible for the effects of gravity. The problem is that all efforts to find the graviton have failed. This is an active area of research, and work continues to find the graviton, and to develop a quantum gravity theory. If we assume gravity mediates through a particle, the case is easily made via Einstein's mass-energy equivalence equations ($E = mc^2$) that all mass, as well as the fundamental forces, reduces to energy quantums.

Although, we are unable to prove conclusively that all masses, including the fundamental forces, consists of discrete energy packets, numerous scientists believe they are. This realization caused Albert Einstein great distress. He wrote in 1954, one year prior to his death, "I consider it quite possible that physics cannot be based on the field concept, i.e., on continuous structures. In that case, nothing remains of my entire castle in the air, gravitation theory included, [and of] the rest of modern physics." Einstein, who grew up in the world of classical physics, was a product of his time. Classical physics utilizes the concept of fields to explain physical behavior. The fields of classical physics are a type of invisible force that influences physical behavior. For example, classical physics explains the repulsion of two positively charged particles due to an invisible repulsive field between them. Modern physics explains this repulsion due to the mediation of photons, which act as force carriers. The main point is that mass and the fundamental forces are ultimately reducible to discrete elements, which equate to discrete packets of energy (quantums).

Space

Next, consider space. In Chapter 1, we discussed the theory that a vacuum, empty space, is like a witch's cauldron bubbling with virtual particles. This theory dates back to Paul Dirac who, in 1930, postulated a vacuum is filled with electron-positron pairs (Dirac sea). Therefore, most quantum physicists would argue that a vacuum is a sea of virtual matter-antimatter particles. This means, even a vacuum (empty space) consists of quantums of energy.

Other forms of energy are in a vacuum. We will illustrate this with a simple question. Do you believe a true void (empty space) exists somewhere in the universe? We can create an excellent vacuum in the laboratory using a well-designed vacuum chamber hooked to state-of-the-art vacuum pumps. We can go deep into outer space. However, regardless of where we go, is it truly void? In addition to virtual particles in empty space, are the gravitational fields. (Viewing gravity as a field is a classical view of gravity. As discussed previously, gravity may mediate via a particle, termed the graviton. For the sake of simplicity, I will use classical phasing, and view gravity as a field.) The gravitational fields would be present in the vacuum chamber, and present even deep in space. Even if the vacuum chamber itself were deep in space, gravitational fields would be present within the chamber. Part of the gravitational field would come from the chamber itself. The rest of the gravitational field would come from the universe. The universe is made up of all types of matter, and the matter radiates a gravitational field infinitely into space. Everything pulls on everything in the universe. The adage, "Nature abhors a vacuum," should read, "Nature abhors a void." Voids do not exist in nature. Within each void is a form of energy. Even if it were possible to remove every particle, the void would contain virtual particles and gravitational fields. As said before, we have not found the graviton, the hypothetical massless particle that mediates gravity, but if you are willing to accept its existence, it is possible to argue that empty space consists of quantums of energy. It bubbles with virtual particles and gravitons.

We can posit another argument that space, itself, is quantized. We will start by asking a question. Is there an irreducible dimension to space similar to the irreducible elements of matter? The short answer is yes. It is the Planck length. We can define the Planck using three fundamental physical constants of the universe, namely the speed of light in a vacuum (c), Planck's constant (h), and the gravitational constant (G). The scientific community views the Planck length as a fundamental of nature. It is approximately equal to 10^{-36} meters (10^{-36} is a one divided by a one with thirty-six zeros after it), smaller than anything we can measure. Physicists debate its meaning, and it remains an active area of theoretical research. Recent scientific thinking is that it is about the length of a "string" in string theory. Quantum physicists argue, based on the Heisenberg uncertainty principle, it is the smallest dimension of length that can theoretically exist.

Does all this argue that space consists of quantized energy? To my mind, it does.

- First, it contains quantized matter-antimatter particles (Dirac sea).
- Second, it contains gravitons (the hypothetical particle of gravity).
- Third, and lastly, space has an irreducible dimension; a finite length termed the Planck length.

Thus far, we have made convincing arguments that mass and space consist of quantized energy. Next, let's turn our attention to time. In Chapter 3, we discussed Planck time (~ 10^{-43} seconds, which is a one divided by a one with forty-three zero after it). As stated in Chapter 3, theoretically, Planck time is the smallest timeframe we will ever be able to measure. In addition, Planck time, similar to the Planck length, is a fundamental feature of reality. We can define Plank time using the fundamental constants of the universe, similar to the methodology to define the Planck length. According to the laws of physics, we would be unable to measure "change" if the time interval were shorter that a Planck interval. In other words, the Planck interval is the shortest interval we humans are able to measure or even comprehend change to occur. This is compelling evidence that time, itself, may consist of quantums, with each quantum equal to a Planck interval. However, this does not make the case that time is quantized energy. To make that case, we will need to revisit the Existence Equation Conjecture discussed in Chapter 12:

$$KE_{X4} = -.3mc^2$$

Where KE_{X4} is the energy associated with an object's movement in time, m is mass and c is the speed of light in a vacuum.

Chapter 12 stated that the Existence Equation Conjecture implied that movement in time (or existence) requires negative energy. The equation, itself, relates energy to the mass (m) that is moving in time. However, just above we argued that all mass is reducible to elementary particles, which ultimately are equivalent to discrete packets of energy via Einstein's mass-energy equivalence equation ($E=mc^2$). This suggests the Existence Equation Conjecture implies that movement in time embodies a quantized energy element. Therefore, if we combine our concept of the Planck interval with the quantized energy nature of time implied by the Existence Equation Conjecture, we can argue that time is a form of quantized energy.

Energy itself

Lastly, one element of reality remains to complete our argument that all reality consists of quantized energy—energy itself. Is all energy reducible to quantums? All data suggests that energy in any form consists of quantums. We already discussed that mass, space, and time are forms of quantized energy. We know, conclusively, that electromagnetic radiation (light) consists of discrete particles (photons). All experimental data at the quantum level (the level of atoms and subatomic particles) tells us that energy exists as discrete quantums. As we discussed before, the macro level is the sum of all elements at the micro level. Therefore, a strong case can be made that all energy consists of discrete quantums.

If you are willing to accept that all reality (mass, space, time, and energy) is composed of discrete energy quantums, we can argue we live in a Quantum Universe. As a side note, I would like to add that this view of the universe is similar to the assertions of string theory, which posits that all reality consists of a one-dimensional vibrating string of energy. I intentionally chose not to entangle the concept of a Quantum Universe with string theory. If you will pardon the metaphor, string theory is tangled in numerous interpretations and philosophical arguments. No scientific consensus says that string theory is valid, though numerous prominent physicists believe it is. For these reasons, I chose to build the concept of a Quantum Universe separate from string theory, although the two theories appear conceptually compatible.

A Quantum Universe may be a difficult theory to accept. We do not typically experience the universe as being an immense system of discrete packets of energy. Light appears continuous to our senses. Our electric lamp does not appear to flicker each time an electron goes through the wire. The book you are holding to read these words appears solid. We cannot feel the atoms that form the book. This makes it difficult to understand that the entire universe consists of quantized energy. Here is a simple framework to think about it. When we watch a motion picture, each frame in the film is slightly different from the last. When we play them at the right speed, about twenty-four frames per second, we see, and our brains process continuous movement. However, is it? No. It appears to be continuous because we cannot see the frame-to-frame changes.

If we have a Quantum Universe, we should be able to use quantum mechanics to describe it. However, we are unable to apply quantum mechanics beyond the atomic and subatomic level. Even though quantum mechanics is a highly successful theory when applied at the atomic and subatomic level, it simply does not work at the macro level. The macro level is the level we experience every day, and the level in which the observable universe operates. Why are we unable to use quantum mechanics to describe and predict phenomena at the macro level?

Quantum mechanics deals in statistical probabilities. For example, quantum mechanics statistically predicts an electron's position in an atom. However, macro mechanics (theories like Newtonian mechanics, and the general theory of relativity) are deterministic, and at the macro level provide a single answer for the position of an object. In fact, the two most successful theories in science, quantum mechanics and general relativity, are incompatible. For this reason, Einstein never warmed up to quantum mechanics, saying, [I can't accept quantum mechanics because] "I like to think the moon is there even if I am not looking at it." In other words, Einstein wanted the moon's position to be predictable, and not deal in probabilities of where it might be.

Numerous scientists, including Einstein, argue that the probabilistic aspect of quantum mechanics suggests something is wrong with the theory. Aside from the irrefutable fact that quantum mechanics works, and mathematically predicts reality at the atomic and subatomic level, it is counterintuitive. Is the probabilistic nature of quantum mechanics a proper interpretation? Numerous philosophical answers to this question exist. One of the most interesting is the well-known thought experiment "Schrödinger's cat," devised by Austrian physicist Erwin Schrödinger in 1935. It was intended to put an end to the debate by demonstrating the absurdity of quantum mechanic's probabilistic nature. It goes something like this: Schrödinger proposed a scenario with a cat in a sealed box. The cat's life or death is depended on its state (this is a thought experiment, so go with the flow). Schrödinger asserts the Copenhagen interpretation, as developed by Niels Bohr, Werner Heisenberg, and others over a three-year period (1924–27), implies that until we open the box, the cat remains both alive and dead (to the universe outside the box). When we open the box, per the Copenhagen interpretation, the cat is alive or dead. It assumes one state or the other. This did not make much sense to Schrödinger, who did not wish to promote the idea of dead-and-alive cats as a serious possibility. As mentioned above, it went against the grain of Einstein, who disliked quantum mechanics because of the ambiguous statistical nature of the science. Einstein was a determinist as was Schrodinger. He felt that this thought experiment would be a deathblow to the probabilistic interpretation of quantum mechanics, since it illustrates quantum mechanics is counterintuitive. He intended it as a critique of the Copenhagen interpretation (the prevailing orthodoxy in 1935 and today). However, far from ending the debate, physicists use it as a way of illustrating and comparing the particular features, strengths, and weaknesses of each theory (macro mechanics versus quantum mechanics).

Over time, the scientific community had become comfortable with both macro mechanics and quantum mechanics. They appeared to accept that they were dealing with two different and disconnected worlds. Therefore, two different theories were needed. This appeared to them as a fact of reality. However, that view was soon about to change. The scientific community was about to discover but one reality exists. The two worlds, the macro level and the quantum level, were about to become one. This tipping point occurred in 2009-2010.

Before we go into the details, think about the implications and questions this raises.

- Do macroscopic objects have a particle-wave duality, as assumed by quantum mechanics at the atomic and subatomic level?
- Can macroscopic objects be modeled using wave equations, like the Schrödinger equation?
- Will macroscopic reality behave similar to microscopic reality? For example, will it be possible to be in two places at the same time?

To approach an answer, consider what happened in 2009.

Our story starts out with Dr. Markus Aspelmeyer, an Austrian quantum physicist, who performed an experiment in 2009 between a photon and a micromechanical resonator, which is a micromechanical system typically created in an integrated circuit. The micromechanical resonator can resonate, moving up and down much like a plucked guitar string. The intriguing part is Dr. Aspelmeyer was able to establish an interaction between a photon and a micromechanical resonator, creating "strong" coupling. This is a convincing and noticeable interaction. This means he was able to transfer quantum effects to the macroscopic world. This is a first in recorded history: we observed the quantum world in order to communicate with the macro world.

In 2010, Andrew Cleland and John Martinis at the University of California (UC), Santa Barbara, working with Ph.D. student Aaron O'Connell, became the first team to experimentally induce and measure a quantum effect in the motion of a human-made object. They demonstrated that it is possible to achieve quantum entanglement at the macro level. This means that a change in the physical state of one element transmits immediately to the other.

> For example, when two particles are quantum mechanically entangled, which means they have interacted and an invisible bond exists between them, changing the physical state of one particle immediately changes the physical state of the other, even when the particles are a significant distance apart. Einstein called quantum entanglement, "spukhafte Fernwirkung," or "spooky action at a distance." Therefore, the quantum level and the macro

> level, given the appropriate physical circumstances, appear to follow the same laws. In this case, they were able to predict the behavior of the object using quantum mechanics. Science and AAAS (the publisher of Science Careers) voted the work, released in March 2010, as the 2010 Breakthrough of the Year, "in recognition of the conceptual ground their experiment breaks, the ingenuity behind it and its many potential applications."

It appears only one reality exists, even though historically, physical measurements and theories pointed to two. The macro level and quantum level became one reality in the above experiment. It is likely our theories, like quantum mechanics and general relativity, need refinement. Perhaps, we need a new theory that will apply to both the quantum level and the macro level.

This completes our picture of a Quantum Universe. We do not know or understand much. Even though we can make cogent arguments that all reality consists of quantized energy, we do not have consensus on a single theory to describe it. When we examine the micro level, as well as the atomic and subatomic level, we are able to describe and predict behavior using quantum mechanics. However, in general, we are unable to extend quantum mechanics to the macro level, the level we observe the universe in which we live. We ask why, and we do not have an answer. Recent experiments indicate that the micro level (quantum level) influences the macro level. They appear connected. Based on all observations, the macro level appears to be the sum of everything that exists at the micro level. I submit for your consideration that there is one reality, and that reality is a Quantum Universe.

We are pressing on to our next and final mystery. Our final mystery, in this section, has to do with the fate of the universe. We know it had a beginning. We know it is evolving, with the expansion of the universe accelerating, and distant galaxies moving away from us faster than the speed of light. What happens when everything runs its course? This is the subject of the next chapter, "How Is the Universe Going to End?"

Fire and Ice

Some say the world will end in fire,
Some say in ice.
From what I've tasted of desire
I hold with those who favor fire.
But if it had to perish twice,
I think I know enough of hate
To say that for destruction ice
Is also great And would suffice.

Robert Frost (1874-1963)

CHAPTER 18

How Is the Universe Going to End?

As Robert Frost so eloquently put it in his poem, *Fire and Ice*, "Some say the world will end in fire, some say in ice." Frost did a marvelous job framing the debate poetically. What does science say?

At the begging of the Twentieth Century, almost every scientist believed the universe was eternal. That is to say, the universe always was and always will be—it is static. In the context of an eternal universe, questions about a beginning or an ending are meaningless. By definition, an eternal universe has no beginning, and it will have no ending. This is what they taught our grandparents as schoolchildren. Overall, the eternal universe found acceptance in both science and religion. Science proclaimed that the universe simply existed, with no evidence to the contrary. Religious leaders simply proclaimed God made the universe, which seems to imply the universe had a beginning. However, since science had no evidence to the contrary, science and religion did not butt heads over this. At the turn of the Twentieth Century, science and religion appeared content with their assertions of the universe. Poetically, you might say all was well in heaven and on Earth.

A little over eighty years ago, our cosmic bubble of an eternal universe was shattered. In 1929, Edwin Hubble discovered that extremely distant galaxies are moving away from us. Indeed, he discovered the farther away the galaxy, the higher the apparent velocity it is moving away from us. Therefore, a galaxy twice as far from us is moving away at twice the speed of a galaxy half the distance from us. Hubble noted that the universe was expanding in all directions. This was a profound discovery that caught the greatest scientific minds of the time, including Einstein, off guard. Prior to Hubble's discovery, the prevalent theory held by the scientific community was that the universe was in a steady state, not expanding or contracting. Even though the evidence was mounting before Hubble conclusively proved the universe was expanding, most scientists held strongly to their paradigm of a steady-state universe.

Surprisingly, Hubble was not the first to discover the universe was expanding. In 1912, Vesto Slipher measured the first Doppler shift (the length of a light wave) of spiral galaxies, and discovered that almost all spiral galaxies were receding from Earth.

Let me break down and explain something fundamental about the observations that Slipher and Hubble made:

- When galaxies move farther away from the earth, scientists like Slipher and Hubble observe a "redshift" in the wavelength of light. That is to say, the light they observe is reddish in color. This occurs because the space between the galaxy and the Earth is expanding. The expansion of the space causes the wavelength of light to "stretch," causing a shift in the color spectrum to the red side, since red has the longest wavelength of any visible light.
- Unfortunately, not much attention was paid to Slipher's findings. Slipher himself did not understand the implications of his discovery.
- In addition, telescopes in 1912 were relatively poor quality, and the nature of what he was measuring was not clearly understood as spiral galaxies. In fact, the term that was used to describe spiral galaxies in 1912 was "spiral nebula" (an indistinct bright patch).
- In addition to ignoring Slipher's findings, they did not recognize the importance of work performed by Alexander Friedman, a Russian cosmologist

and mathematician, who, in 1922, used Einstein's equations of general relativity to demonstrate that the universe was expanding in contrast to the static universe model advocated by Einstein.

- In 1924, Hubble, using the 100-inch Hooker telescope at the Mount Wilson Observatory, confirmed that the "spiral nebula" were indeed other galaxies, and was able to estimate distances to those galaxies whose redshifts had already been measured by Slipher.
- In 1927, Georges Lemaître, a Belgian physicist and Roman Catholic priest, independently derived Friedmann's equations, and proposed that the inferred recession of the galaxies was due to the expansion of the universe.
- In 1929, Hubble discovered a correlation between distance and recession velocity. This became known as Hubble's law, namely the farther away the galaxy, the higher the apparent velocity it had away from the earth.
- Even the mountains of evidence leading to Hubble's discoveries were initially met with resistance. However, other astronomers confirmed Hubble's measurements, and the evidence was slowly becoming overwhelming.

In light of all the discoveries delineated above, Einstein abandoned his static universe theory, and termed his assumption of a "cosmological constant" his greatest blunder. Einstein constructed the "cosmological constant" to force his theory of relativity to predict a static universe. At this point, Hubble's discovery was gaining traction in the scientific community.

In 1931, Georges Lemaître, a Belgian priest, astronomer and professor of physics at the Catholic University of Leuven, suggested that if the expansion of the universe were projected back in time, all the mass of the universe would be concentrated into a single point, which he called a "primeval atom." Lemaître believed that it was at this point that the fabric of time and space came into existence. Most scientists today agree with him. However, Lemaître did not coin the phrase "Big Bang." It was English astrophysicist Fred Hoyle in 1950. Hoyle, who championed a rival cosmological theory, wanted to discredit or at least cast doubt on the expansion-of-the-universe theory. He intended the term "Big Bang" to be a term of derision. To his, and everyone else's surprise, it caught on. I think that is unfortunate because the term suggests a colossal explosion. In reality, to the best of our current measurement capabilities, nothing exploded. Energy expanded. However, scientists continued to argue the correctness of the Big Bang model until 1964.

In 1964, Arno Penzias and Robert Wilson discovered the cosmic microwave background radiation. At the time, they were conducting diagnostic observations using a new microwave receiver owned by Bell Laboratories. Why did this turn out to be crucial? To address this question, we will need to discuss the cosmic microwave background.

Cosmic microwave background radiation is the thermal radiation believed left over from an early stage in the development of the universe. At the instant the Big Bang occurred (consider "instant" to mean several days, when we speak cosmologically), the early universe was in thermal equilibrium. Photons (packets of energy or commonly referred to as light) were being continually emitted and absorbed, with energies over the entire spectrum of a perfect black body. The emission and absorption of photons continued until the universe expanded further and, because of the expansion, cooled. In effect, the energy of the universe became less dense via the expansion, and the universe became too cold to sustain the photon emission and absorption. At this point, the emission and absorption of photons stopped. However, the universe was still hot enough for electrons and protons to remain unbound (atoms like hydrogen had not yet formed). When a photon collided with an electron, it was not absorbed. It was reflected, scattering light diffusely in all directions, and causing the early universe to be opaque. As the universe continued to cool, atoms like hydrogen began to form, causing free electrons to become scarce. Since photons rarely scatter from neutral atoms, like hydrogen, the radiation decoupled from matter. The photons that remained make up the cosmic wave background radiation. The energy of photons was red-shifted by the expansion of the Universe. This preserved the blackbody spectrum, but caused its temperature to fall. This caused the left-over photons (the cosmic microwave background radiation) to fall into the microwave region of the electromagnetic spectrum, which is invisible to our human eyes, but detectible by sensitive radio telescopes.

Science estimates that the cosmic wave background radiation formed about 379,000 years after the Big Bang. In cosmic terms, this length of time is a mere "blink of the eye," since the universe is about 13.7 billion (13,700,000,000) years old. This mathematically works out to .003% of the total age of the universe. In other words, it occurred extremely early in the formation of our universe.

Arno Penzias and Robert Wilson's discovery of the cosmic microwave background in 1964 tipped the balance of scientific opinion in favor of the Big Bang theory. In 1978, Penzias and Wilson were awarded the Nobel Prize for their discovery. To show how pivotal this discovery was, I will quote the famous physicist and cosmologist, Stephen Hawking. In 1965, one year after the discovery of the cosmic microwave background radiation, Stephen Hawking stated this discovery put "the final nail in the coffin of the steady-state theory."

Starting with Hubble's discovery of an expanding universe in 1929, within thirty-five years, most of the scientific community did a complete reversal, turning their backs on the eternal universe theory, and embracing the Big Bang. In addition, the Catholic Church quickly embraced the Big Bang theory, since it appeared to be in concert with scripture, namely the universe had a beginning. Once again, you might say all was well in heaven and on Earth.

As scientists began to think about an expanding universe, they reasoned that eventually gravity would play into the equation, halt the expansion, and even reverse it. In other words, up to 2008, mainstream science believed that the expansion of the Big Bang would eventually be slowed by gravity, then halted, and gravity would pull everything back together in what science termed a "Big Crunch." This is a reasonable belief. If you throw a rock straight up in the air, its velocity is eventually slowed by gravity. At a point, the rock thrown upward stops in midair, and gravity pulls it back to your hand. If you threw it straight, it would return exactly to your hand. However, as we discussed in the Introduction, when the expansion of the universe was measured in 1998, by Saul Perlmutter, Brian P. Schmidt, and Adam G. Riess, a startling discovery was made. The expansion was not slowing down. It was accelerating. Gravity did not appear to be playing a prominent role. In fact, a new and unknown force, termed "dark energy," seemed to be in charge. This new force, dark energy, is still a mystery (see Chapter 10).

We have irrefutable evidence that the universe had a beginning—the Big Bang—and irrefutable evidence that the expansion of the universe is accelerating. What does this imply?

Based on all known data, the accelerated expansion of the universe implies that eventually all galaxies will move away from us to the point they are beyond our cosmological horizon. We will no longer be able to see them. All evidence of the Big Bang will have disappeared along with the galaxies. The universe will consist of the Milky Way galaxy.

When is this all going to happen? No one knows. Most scientists agree on periods of billions of years, but no one knows the exact number of billions. Some theories calculate the end of the universe, but they hold little sway in mainstream science. All we certainly know is that the universe is 13.7 billion years old, which suggests a change on a cosmological scale moves slowly. The end is likely billions of years in the future.

Will human kind survive to witness the end? We will discuss the answer to this question fully in Chapter 19. Here's a glimpse into our future. If we survive, it will not be on Earth. Our sun, which is about five billion years old, is halfway through its life. In another five billion years, it will die—and destroy the Earth as it dies. If we survive, we will become nomads, or we will live on another Earth-like planet, perhaps even on a planet in a parallel universe.

Unfortunately, we know how Robert Frost's poem, *Fire and Ice,* ends. All the big money is on "ice." If you feel down and lonely after reading this chapter, I think reading Section III, "Are We Alone?" will cheer you up a bit.

Section III

Are We Alone?

"Our sun is one of 100 billion stars in our galaxy. Our galaxy is one of billions of galaxies populating the universe. It would be the height of presumption to think that we are the only living things in that enormous immensity."

Wernher von Braun (1912-1977)

"Using the most conservative methods at his disposal, [Isaac] Asimov attempted to come up with a plausible number of habitable planets amongst the estimated 300 billion stars in the Milky Way—while also trying to calculate the number that might have been home to civilizations of alien life at or around our own current level of biological evolution. The number he came up with? 500,000."

Eric Pfeiffer, The Sideshow – Tue, Dec 6, 2011

CHAPTER 19

Is There Another Earth?

How probable is it to find another Earth-like planet? No one even knew there were any planets outside our solar system until relatively recently. It is hard to discover a distant planet since they are small and extremely dim, compared to a star.

Our first confirmed discovery of another planet outside of our solar system occurred in 1992. Aleksander Wolszczan and

Dale Frail announced the discovery of two planets orbiting a pulsar (Wolszczan, A.; Frail, D. A. (1992). "A planetary system around the millisecond pulsar PSR1257+12." Nature 355 (6356): 145-147). As is often the case in science, one discovery leads to another, and slowly more planets were discovered.

The discovery of new planets around other stars ("exoplanets") increased dramatically with the March 2009 launch of NASA's Kepler spacecraft. An important part of its mission was to find Earth-like planets. As of this writing, the Kepler spacecraft, with the associated Kepler Space Telescope team, has accelerated the discovery of new exoplanets, bringing the total to 760 known extrasolar planets listed in the Extrasolar Planets Encyclopedia. The team also claims to have found the first Earth-like planet.

Before we proceed, we'll address a fundamental question. What makes a planet Earth-like? When we use the term "Earth-like," we mean the planet resembles the Earth in three crucial ways:

1) It has to be in an orbit around a star that enables the planet to retain liquid water on one or more portions of its surface. Cosmologists call this type of orbit the "habitable zone." Liquid water, as opposed to ice or vapor, is crucial to all life on Earth. There might be other forms of life significantly different from what we experience on Earth. However, for our definition of an Earth-like planet, we are confining ourselves to the type of life that we experience on Earth.

2) Its surface temperature must not be too hot or too cold. If it is too hot, the water boils off. If it is too cold, the water turns to ice.
3) Lastly, the planet must be large enough for its gravity to hold an atmosphere. Otherwise, the water will eventually evaporate into space.

In December 2011, Kepler astronomers announced the discovery of an Earth-like planet, now called "Kepler 22b." It is about 2.4 times wider than the Earth, and circles a star that is similar to our sun. They estimate Kepler 22b's average surface temperature to be about 72°F (degrees Fahrenheit). It is 600 light years from Earth, which cosmologically speaking makes it a near neighbor. The most crucial aspect that makes the planet Earth-like is that it is in the habitable zone.

We have the answer to the question: is there another Earth-like planet? The answer is yes, and it is relatively close, by galactic standards. These are truly exciting times to be alive. When I started the research for this book, we did not know about Kepler 22b. I would have had to make an argument on faith that somewhere out in the universe there had to be another Earth-like planet. The probabilities would have supported that argument. In light of this new discovery, I do not need to argue or ask for faith. Kepler 22b is, to the best of our scientific measurements, Earth-like. Perhaps when our grandchildren's grandchildren read this book or one like it, it will be old hat. We will have discovered countless Earth-like planets, and perhaps our grandchildren's grandchildren will be living on one of them.

If it is Earth-like, will it have life on it? The odds are it will. Hard to believe? It will become more believable if we examine how life spreads around in the universe. To understand this phenomenon, we will start with our own planet, which we know had life on it when the dinosaurs became extinct 65 million years ago.

From the fossil record, the extinction of the dinosaurs most likely occurred when an asteroid, approximately 10 km in diameter (about six miles wide), and weighing more than a trillion tons, hit Earth. The impact killed all surface life in its vicinity, and covered the Earth with super-heated ash clouds. Eventually, those clouds spelled doom for most life on the Earth's surface. However, this sounds like the end of life, not the beginning. It was the end of life for numerous species on Earth, like the dinosaurs. However, the asteroid impact did one other incredible thing. It ejected billions of tons of earth and water into space. Locked within the earth and water—was life. The asteroid's impact launched life-bearing material into space. Consider this a form of cosmic seeding, similar to the way winds on Earth carry seeds to other locations to spread life.

Where did all this life-bearing earth and water go? A scientific paper from Tetsuya Hara and colleagues, Kyoto Sangyo University in Japan, (*Transfer of Life-Bearing Meteorites from Earth to Other Planets*, Journal of Cosmology, 2010, Vol 7, 1731-1742), provide an insightful answer to our question. Their estimate is that the ejected material spread throughout a significant portion of the galaxy. Of course, a substantial amount of material is going to end up on the Moon,

Mars, and other planets close to us. However, the surprising part is that they calculate that a significant portion of the material landed on the Jovian moon Europa, the Saturnian moon Enceladus, and even Earth-like exoplanets. It is even possible that a portion of the ejected material landed on a comet, which in turn took it for a cosmic ride throughout the galaxy. If any life forms within the material survived the relatively short journey to any of the moons and planets in our own solar system, the survivors would have had over 64 million years to germinate and evolve.

Would the life forms survive an interstellar journey? No one knows. Here, though, are incredible facts about seeds. The United States National Center for Genetic Resources Preservation has stored seeds, dry and frozen, for over forty years. They claim that the seeds are still viable, and will germinate under the right conditions. The temperature in space, absent a heat source like a star, is extremely cold. Let me be clear on this point. Space itself has no temperature. Objects in space have a temperature due to their proximity to an energy source. The cosmic microwave background, the farthest-away entity we can see in space, is about 3 degrees Kelvin. The Kelvin temperature scale is often used in science, since 0 degrees Kelvin represents the total absence of heat energy. The Kelvin temperature scale can be converted to the more familiar Fahrenheit temperature scale, as illustrated in the following. An isolated thermometer, light years from the cosmic microwave background, would likely cool to a couple of degrees above Kelvin. Water freezes at 273 degrees Kelvin, which, for reference, is equivalent to 32 degrees Fahrenheit.

Once the material escapes our solar system, expect it to become cold to the point of freezing. If the material landed on a comet, the life forms could have gone into hibernation, at whatever temperature exists on the comet. If an object in space passes close to radiation (such as sunlight), its temperature can soar hundreds of degrees Kelvin. Water boils at 373 degrees Kelvin, which is equivalent to 212 degrees Fahrenheit. We have no idea how long life-bearing material could survive in such conditions. However, our study of life in Earth's most extreme environments demonstrates that life, like Pompeii worms that live at temperatures 176 degrees Fahrenheit, is highly adaptable. We know that forms of life, lichens, found in Earth's most extreme environments, are capable of surviving on Mars. This was experimentally proven by using the Mars Simulation Laboratory (MSL) maintained by the German Aerospace Center. It is even possible that the Earth itself was seeded via interstellar material from another planet. Our galaxy is ten billion years old. Dr. Hara and colleagues estimate that if life formed on a planet in our galaxy when it was extremely young, an asteroid's impact on such a planet could have seeded the Earth about 4.6 billion years ago.

Would any of the life-bearing material be able to reach Kepler 22b? The trip to Kepler 22b would have taken an Earth meteorite about 30 million years to reach it. However, the amount of material reaching Kepler 22b would likely be small, due to dispersion. To understand dispersion, consider a flashlight. If you shine the light on a nearby wall, you will see a bright spot on the wall. This is due to the high number of photons that concentrate on the wall to form the bright spot. However, if you move farther away from the wall, the bright spot becomes larger and dimmer. The photons are spreading over a larger area, and are not as concentrated. If you move back far enough, the bright spot will eventually fade, and only a faint glow will be seen on the wall. This phenomenon is called dispersion. The photons being emitted from the flashlight spread apart and become less dense the farther they travel from the flashlight. This same phenomenon occurred when the dinosaur-killing asteroid ejected material from the Earth. As it traveled farther from the Earth, the ejected material began to spread further apart (disperse). Even if a portion of life-bearing material made it to Kepler 22b, the smaller meteorites may have simply burned up in its atmosphere. This is what happens on Earth. Since Kepler 22b is twice the diameter of Earth, it is likely to have a dense atmosphere. Yet, the possibility of seeding Kepler 22b with Earth's life-bearing material is still possible. If it happened, the life forms would have had 35 million years to evolve.

This is essentially a new way of thinking about the origin of life on Earth, and on other planets. Once life forms on a planet, it appears that the cosmos itself takes care of spreading it throughout the galaxy. Therefore, you may begin to conclude that life on other planets would look a lot like life on Earth. That would be unlikely, unless the planet closely resembled Earth. As we see when we study life in extreme environments on Earth, life adapts to the environment. Therefore, on a large planet where gravity might be three times greater than on Earth, the life forms would have evolved to accommodate the increased gravity. Perhaps they would be closer to the ground, and have larger legs or even no legs, like snakes. Perhaps they have larger eyes if the planet has low light. Perhaps they have no eyes, like worms, if the planet is in darkness. Science fiction writers do an excellent job of conjuring up extraterrestrial life based on the planet from which the life forms originate. You can use your imagination to draw your own conclusions on what they might look like, based on their planet of origin.

All the evidence suggests, Earth-like planets are in our own galaxy. We have already discovered one, Kepler 22b. In the coming years, we will likely discover more. The evidence suggests that we are going to find life on the Earth-like planets. We may even find life on planets that are not Earth-like. However, are you wondering: will we find advanced aliens? Not simply life, but intelligent life? This concept has been the central theme of science fiction for well over a century, and is the subject of our next chapter.

"Could it be that God was an extra-terrestrial? What do we mean when we say that heaven is in the clouds? From Jesus Christ to Elvis Presley, every culture tells us of high-flying bird men who zoom around the world creating magnificent works of art and choosing willing followers to share in the eternal glory from beyond the stars. Can all these related phenomena merely be dismissed as coincidence?"

Erich von Däniken, *Chariots of the Gods*

CHAPTER 20

Are There Advanced Aliens?

Most prominent scientists will not weigh in on this subject for fear of losing their credibility. I highly applaud Stephen Hawking for his courage. He put his reputation on the line, when he warned against attempting to contact extraterrestrials. No scientist of his eminence had come out strongly warning humankind from making contact with advanced aliens. What exactly did the most prominent cosmologist of our time say?

In a Discovery Channel documentary, "Into the Universe with Stephen Hawking," which aired in April 2011, Stephen Hawking warned against attempting to contact aliens. He believes it is likely that advanced aliens exist, and that contacting them would be a threat to the Earth. His exact words, "If aliens visit us, the outcome would be much as when Columbus landed in America, which didn't turn out well for the Native Americans. We only have to look at ourselves to understand intelligent life might develop into something we wouldn't want to meet." He further speculated that the capabilities of advanced aliens "would be only limited by how much power they could harness and control, and that could be far more than we might first imagine." One example he gave was that the aliens might be able to harness the total energy of a star. He painted a bleak scenario: "Such advanced aliens would perhaps become nomads, looking to conquer and colonize whatever planets they can reach." To Dr. Hawking, it was obvious that aliens exist, "To my mathematical brain, the numbers alone make thinking about aliens perfectly rational." He added, "The real challenge is to work out what aliens might actually be like."

It is literally impossible for a cosmologist and physicist of Stephen Hawking's stature to make assertions about aliens, and not have scientists all over the world weigh in on his assertions. His remarks received immediate reaction, both pro and con. It divided the scientific community, with a number of scientists backing Dr. Hawking's warning, and others dismissing it. The Search for Extraterrestrial Life Institute (SETI) was one of the first to comment. According ABC World News, Jill Tarter, director of the Center for SETI Research, said that SETI only listens for signals of technology from extraterrestrial life. SETI does not broadcast into space.

Most of the scientific community considers the SETI project legitimate scientific research, since they employ scientifically accepted radio-receiver methodologies to search for evidence of extraterrestrial life. No similar scientific consensus exists regarding the legitimacy of UFOlogy, which is the study of UFOs (unidentified flying objects). However, it is essential not to equate the legitimacy afforded the SETI project's research as proof that advanced aliens exist. Nor should we cast doubt on the existence of advanced aliens based on the skepticism that surrounds UFOlogy.

SETI has been listening for about 50 years and, to date they have not detected any signals that would suggest alien technology. However, it a vast universe, and our ability to detect the alien technology signals may not be up to the task at this time. Therefore, even if the aliens have broadcasted, we may not have the right receiver.

Other broadcasts have been sent from Earth into space, which started with our first radio broadcast about one hundred years ago. However, our standard radio and television broadcasts are relatively weak, and diffuse rapidly in the vastness of space. At best, even if our faint signals were detectable, they would have traveled about 100 light years from Earth. If you consider that our galaxy, the Milky Way, is 100,000 light years across, we have not reached out far at this point. Perhaps, considering Dr. Hawking's warning, that is in our favor.

In the last chapter, we discussed a recent discovery of an Earth-like planet orbiting a star similar to our own. However, it is a long way from finding a planet that may support life—to finding life. We have not found extraterrestrial life. Even if we do, it is an even longer way from finding life to finding intelligent life. Why is the greatest cosmologist in the world, Stephen Hawking, warning us to keep a low profile? Because the odds strongly favor that life exists, and even intelligent life exists somewhere else in the universe. Based on discoveries made using the Kepler spacecraft, we know that the Milky Way has a minimum of 100 billion planets. That means on average at least one planet per star, and approximately 1,500 planets, are within 50 light-years of Earth. Given the vastness of the universe, the odds strongly favor life and intelligent life. Again given the vastness of space, there are probably more Earth-like planets in Earth-like solar systems, with large Jupiter-like planets that are in a similar position to our own Jupiter. Our Jupiter acts as a cosmic magnet for asteroids. This helps protect us from cosmic calamities such a large asteroid collision with Earth. Our entire solar system, given our position in orbit around our sun—and having a large planet like Jupiter in the right orbit to help protect us from asteroid collisions—can be described as a "Goldilocks" solar system. However, given the vastness of space, it is entirely possible for this type of solar system to have evolved numerous times.

If intelligent life exists, imagine if they evolved one million years earlier than we did. From a cosmological perspective, one million years is a blink of an eye. Imagine what our capabilities will be a thousand years in the future, assuming humankind exists one thousand years in the future. It is entirely reasonable to assume intelligent life may have gotten an earlier start in the universe, and be scientifically more advanced.

This brings us to the Fermi paradox, which poses a deceptively simple question: if the probability of advanced aliens is so high, why haven't we detected them or been contacted by them? The paradox has to do with the high probability of existence, in this case advanced aliens, and the lack of evidence.

In 1950, employee Enrico Fermi was walking to lunch with his colleagues at Los Alamos National Laboratory. The topic of UFOs came up because of numerous sightings and reports sensationalized by the media. Although the conversation started on a light note, it soon became serious. Fermi and his colleagues began to discuss the possibility of faster-than-light travel, which from Einstein's special theory of relativity, is impossible. However, if advanced aliens were going to visit the Earth, they would likely need to travel faster than light given the vast distances between interstellar destinations. Although Fermi's colleagues considered faster-than-light travel a long shot, Fermi believed that science would discover a way to make objects travel faster than light within a decade. He was wrong about that, but his main point was a question. In the middle of lunch, he jumped up and asked, "Where is everybody?" His point, if the universe contains advanced extraterrestrial life, where is the evidence? Fermi began to calculate the potential existence of advanced aliens. His rough calculations indicated that the Earth would have been visited numerous times, from ancient times to the present. This became known as the Fermi Paradox, namely the probability that advanced aliens exist does not square with the lack of evidence. Where are they? Insight to address this question is one page away, in our next chapter, "The Search for Extraterrestrial Intelligence."

"[The octopus has] an amazing skin, because there are up to 20 million of these chromatophore pigment cells and to control 20 million of anything is going to take a lot of processing power. ...These animals have extraordinarily large, complicated brains to make all this work. ...And what does this mean about the universe and other intelligent life? ....I would expect, personally, a lot of diversity and a lot of complicated structures. It may not look like us, but my personal view is that there is intelligent life out there."

— Roger T. Hanlon, From transcript of PBS TV program Nova episode 'Origins: Where are the Aliens?' (2004).

CHAPTER 21

The Search for Extraterrestrial Intelligence

The Fermi Paradox, discussed in the last chapter, galvanized a number of scientists to ask similar questions.

- Is the probability of finding extraterrestrial intelligent life in the universe high, or is it rare?
- Have we encountered any evidence of extraterrestrial intelligent life?

In 1961, Dr. Frank Drake, an American astronomer, and a founder of SETI (search for extraterrestrial intelligence), formulated an equation known as the Drake equation, to calculate the number of intelligent civilizations in our Milky Way galaxy. It had, however, several serious drawbacks. First, the equation had at least four utterly unknown terms in it, namely 1) the fraction of planets with life, 2) the odds life becomes intelligent, 3) the odds intelligent life becomes detectable, and 4) the detectable lifetime of civilizations. It suffered from a highly questionable premise, namely that advanced alien civilizations arise and die out in their own solar system. Therefore, scientists like Dr. Carl Sagan could optimistically predict over one million advanced alien civilizations in 1966, while other less-optimistic scientists predicted we were alone. All used the same equation, but with different assumptions for the unknowns. As you can imagine, instead of resolving the paradox, it fueled it. In fairness though, the Drake Equation was not proposed as a hypothesis. It was not intended to be proved or disproved. Its main purpose was to fire our imaginations to the possibility that extraterrestrial life may exist in our galaxy. The Drake equation served to create the agenda of the first SETI meeting.

When you consider the scale of our galaxy, the Milky Way, current estimates are over 100 billion stars, with each star having at least one planet. Even if intelligent life were rare, the scale of our galaxy would argue it must exist somewhere. If there was one other, and they had a fifty million year head start, which is small by cosmic standards, they would have had time to colonize the entire galaxy by now, even using spaceships only capable of sub-light speed. Consider our own history. First, we explored the Earth, our own relatively small planet. As technology permitted, we explored our moon and planets in our own solar system. We have even launched satellites, Voyagers 1 and 2, beyond our solar system. The last images we have of them were taken on Valentine's Day 1990 as they headed into deep space. Our instinctive need to explore is more than simple curiosity. It is essential to our survival, to be discussed in the next chapter. If we use ourselves as an example, advanced aliens would likely share our instinctive need to explore. As resources became scarce on their own planet, they would look to other planets. If their sun began to die, they would migrate to a new planet with a more suitable habitat. Science has a principle for this model—the "mediocrity principle," which states nothing is unusual about the Earth's evolution. Therefore, our own galaxy should be host to numerous advanced alien civilizations. This same argument can be extended to the billions of other galaxies in our universe. We have discovered numerous galaxies similar to ours. In fact, current estimates state 70 sextillion (a seven with twenty-two zeros after it) stars are in the visible universe. Therefore, the odds of advanced aliens existing in the universe are high. So, where are they?

One way to answer the "where are they?" question is to find evidence. The quest to find evidence of extraterrestrial life started about 1960, and continues to this day. I will go into detail on the scientific thrusts to find extraterrestrial life, but first here's an overview. As of 2012, humankind has been able to visit the moon, and to send robots to Mars, proven by pictures and geological samples. To date, the evidence is inconclusive regarding the existence of life, past or present. We are unable to do interstellar travel, so we are confined to exploring the nearby moons and planets in our own solar system, or observing from afar. That means the life on the planets we are observing from afar would have to change the planets in a way we can detect, or do something we can detect. For example, if they sent radio signals, we might be able to detect them. If the life on the planets is not advanced, we have only a slight chance we would detect it, without taking samples. For example, if life exists under the surface of Mars within the proximity of the Martian polar ice caps, samples taken from those regions may allow us to detect it. In addition, we can examine meteorites for evidence of life. As of 2012, NASA has collected 34 Mars meteorites. Three of them suggest life existed on Mars in the past. The question of life on Mars, past or present, is still an open question. We do not know, but portions of the data appear promising. My main point is that the search for extraterrestrial life is difficult. The following are the significant efforts that have been or are being taken:

Using astronomy to search for advanced aliens to establish proof of their existence—Astronomers have been looking at the heavens since the invention of the telescope in about 1608. The idea here is that astronomers may discover a phenomenon that suggests or requires the existence of intelligent life. For example, consider the Seyfert galaxies. These are a class of galaxies with extremely bright nuclei. Carl Keenan Seyfer discovered them in 1943. Initially the Seyfert galaxies' enormous and directed-energy output was thought to be the result of an industrial accident on a planet inhabited by advanced aliens. However, science now attributes this phenomenon to gravity causing objects to be captured (accretion) by black holes.

Even the SETI researchers, better known for their radio-signal reception work, engage in near-conventional astronomy. I use the word "near" because, in addition to looking at likely planetary systems, they are searching for laser beams from advanced aliens. It seems far out, doesn't it? We would have to be looking at the right place, at the right time, and the laser beam would have to be distinguishable from the star (solar system) we are observing. However, we do not know how advanced aliens would communicate. Conventional radio to them may have become obsolete. If they are extremely advanced and able to make a gamma ray laser, then according John A. Ball (MIT Haystack Observatory), they could transmit a two-millisecond pulse encoding 1×1018 bits of information, which would embody the total information content of Earth, down to the last gene sequence. Interestingly, we observe about one gamma ray burst per day. SETI has removed them from their radio transmission surveillance because the gamma rays do not penetrate our atmosphere.

To date, no conclusive evidence has been recorded of extraterrestrial life via astronomy. However, the potential, though small, still exists that conventional astronomy may detect the existence of advanced aliens (especially if they use flying saucers to visit us).

Searching for radio emissions from advanced aliens to establish proof of their existence—Since our discovery of the radio in 1895, we have been beaming radio transmissions into space. Most scientists believe the invention of the radio and radio telescopes would be a natural technological evolution by any intelligent life. Therefore, it would be reasonable to conclude that advanced aliens may have transmitted proof of their existence. The timeframe of their transmissions would depend on when they evolved. If their evolution were concurrent with ours, their transmissions would have started about a century ago. However, if they evolved millions of years ahead of us, their transmissions could have started millions of years ago.

In fact, the whole notion of listening for radio transmissions from aliens dates back to 1896, when Nikola Tesla promoted the idea that the radio could be used to contact advanced extraterrestrial life. In the early 1900s, Guglielmo Marconi, the inventor of the radio, claimed to have picked up Martian radio signals. Other iconic scientists, like Lord Kelvin, credited with inventing the telegraph, added fuel to the radio search for advanced aliens by publicly stating that the radio represented a possible way to detect and even contact them.

When scientists of the stature of Tesla, Marconi, and Kelvin speak, the world listens. In 1924, Mars was closer to Earth than any time in the last 100 years before or since. Obviously, this would be an excellent time to listen for radio transmissions from Mars. To avoid cluttering the Martian signals with our own, a "National Radio Silence Day" was promoted by the United States. For a 36-hour period, during August 21-23, 1924, all radios were silent for five minutes at the beginning of each hour. Concurrently, a dirigible was used to lift a radio up in order to receive signals 3 kilometers above the United States Naval Observatory. A select few listened, including the chief cryptographer of the U.S. Army, William F. Friedman. No radio transmissions from Mars were reported.

The most famous human enterprise listening for alien radio transmissions is SETI, which is not a single organization, but rather a group of organizations that employ radio technology to search for advanced extraterrestrial life. This includes Harvard University, the University of California, Berkeley, and the SETI Institute. Astronomer Frank Drake, using a small radio telescope, undertook the first SETI experiment in 1960. In 1961, the first SETI conference was held at Green Bank, West Virginia. From this humble beginning, SETI was launched. It is still highly active in its search for extraterrestrial radio transmissions as of this writing.

SETI technology has improved vastly. They are searching

more frequencies than ever before. However, to date we have no confirmable evidence. SETI researchers have intercepted signals twice, once in 1977 and once in 2003, that may have been alien in nature, but they were not able to confirm the results. In fact, after more than five decades of searching, no confirmable radio transmission evidence of advanced aliens exists. However, to be fair to SETI, we need to examine their two greatest obstacles.

1) Scale Problems—The universe is enormous, and SETI has had to confine its search to sun-like solar systems within about 200 light years of Earth. Our galaxy is about 100,000 light years across. This may appear as if they have examined about 20% of our galaxy, but that would be incorrect. They focus on high-probability solar systems (ones similar to our own), and thinly slice space looking for the radio transmission. Therefore, the real number is much less than 20%. If it is viewed in terms of the volume, SETI has covered one-fifteen millionth of our own Milky Way galaxy. This, however, is likely to improve. If we add the recent upgrades that SETI made in 2007, namely the Allen Telescope Array, located in northern California, SETI is able to extend its search radius to 25,000 light years. This enables SETI to examine the 40 billion solar systems closer to the center of our galaxy. Still though, we are looking for a needle in a very large galactic haystack.

2) Technical Hurdles—Our transmitted radio and television signals disperse relatively rapidly in space. They would require extremely sensitive radio telescopes to detect. To understand this, imagine someone holding a candle at night a few feet from you. You are able to see it clearly. This is because numerous photons from the candle are reaching your eyes. Next, imagine that person moves farther away from you. The farther away the person moves, the dimmer the candle becomes. After a while, you will not be able to see the candle at all. The photons of the candle spread out over distance. Initially, when you were close to the candle, numerous photons reached your eyes. As the candle moved farther away, the photons spread out over a larger area, and fewer of them reach your eyes. This is why the candle became dimmer. Eventually, the candle was so far away, too few photons were reaching your eyes for your eyes to sense them.

SETI estimates that even with a sensitive radio telescope, as the extremely large Arecibo Observatory radio telescope in Puerto Rico, the Earth's radio and televisions transmissions would only be detectable at a distance within 0.3 light years. Therefore, unless the advanced aliens used highly directed transmissions, we would likely not detect them. In addition, if the advanced aliens compressed their data, similar to data downloads from the Internet, the compressed data would appear as noise to us. In addition, advanced aliens may be using frequencies we are not monitoring or do not penetrate our atmosphere. The list of technical hurdles is numerous. Their sheer number and complexity has cast doubt on the entire SETI methodology. Critics believe the SETI efforts are futile, since the technical hurdles regarding the interception of advanced alien radio transmissions are enormous.

Searching for and analyzing potential alien artifacts to establish proof of their existence—Similar to the way archaeologists uncover lost civilizations on Earth by analyzing the artifacts left behind, various researchers believe the past presence of advanced aliens could be detected in a similar manner. This is a reasonable approach. It has historically provided evidence of civilizations that appear to have simply vanished. For example, the Mayan calendar is supposedly predicting the end of the world on December 21, 2012. Unfortunately, this is a poor example of a lost civilization, since it never disappeared. In fact, the Maya and their decedents still populate the Maya area, and continue to honor traditions that date back centuries. Millions of Mayans still speak the Mayan language. As for the Maya calendar, most scholars do not interpret it to predict the end of the world.

A real example of a lost civilization can be found in our own North American backyard. The Anasazi lived in the bordering parts of Utah, Arizona, New Mexico, and Colorado. The Anasazi civilization emerged about 1100 BC, and appeared to vanish about 1100 AD. However, did they really vanish? Most archeologist think not. They did abandon their traditional homeland. In a number of cases, the "lost" civilizations are not lost. They move to a different location for reasons that generally relate to survival, like water and food availability. However, the point is that we know about the Anasazi civilization by studying the artifacts lefts behind, including their dwellings, pottery, tools, and the like.

Proponents of ancient alien visits to Earth point to the numerous alien-like artifacts. These include:

- References in religious texts, such as the Book of Ezekiel (Biblical Old Testament)
- Physical evidence such as Nazca Lines, which depict drawings that can only be fully seen from the air (Peru)
- Ancient aircraft-type models, like the Saqqara Bird (1898 excavation of the Pa-di-Imen tomb in Saqqara, Egypt), and small gold model "planes" (Central America and coastal areas of South America)
- Unusual ancient monuments and ruins such as the Giza pyramids in Egypt, Machu Picchu in Peru, Baalbek in Lebanon, the Moai on Easter Island, and Stonehenge in England. Proponents of ancient alien visits argue these structures could not have been built without alien help. They argue that the ability to build them was beyond the capability of humankind at the time they were built.

This is a sampling that proponents of ancient aliens provide as evidence that the Earth has been visited since ancient times by advanced aliens. Numerous books forward this theory. The most famous was written by Erich von Däniken, and published in 1968 (*Chariots of the Gods?)*.

Obviously, this is a speculative theory, and not everyone agrees. In fact, there is considerable disagreement. Several disagree on religious grounds, like the Christian creationist community. Other critics simply say the evidence is subject to various interpretations. In reality, we have not found irrefutable evidence—the "smoking gun." For example, if we found an electromagnetic transmitter (a radio) of unknown origin inside a newly discovered 3,000-year-old pyramid, that would be a smoking gun.

Two primary schools of thought are posited to address the question: where are they?

1) Little-to-no intelligent life exists in the universe, with the exception of us.
2) The universe has intelligent life, but we see no evidence of it.

We will examine each in detail, and start by asking the question: why would there be little-to-no intelligent life in the universe, except us? These are a few of the reasons given:

- **Rare Earth hypothesis**—Proponents maintain the conditions required for intelligent life to form are extremely rare, making the possibility of there being another intelligent civilization "zero." Opponents argue that the universe already has met the conditions of intelligent life once, and that since the universe is immense, the probability of meeting the conditions again are high. They argue that the evolution of life does not need to follow the same evolutionary path as Earth. Given our discussion on cosmic seeding in the last chapter, the opponents appear to have a valid point.

- **Doomsday argument**—Proponent of this theory hold that it is the nature of advanced life to destroy itself and others. Given the way our world has evolved, the proponents of this theory have a plausible argument. I think a case can be made that Stephen Hawking's warning against making contact with advanced aliens falls under the Doomsday argument. I think elements of this argument are valid. Advanced aliens would likely have faced the same challenges the human race faces. Indeed, they might have suffered internal and even interstellar wars to the point that their civilizations were destroyed, or bombed back to the Stone Age (non-technically advanced). On the other hand, we are still here. In addition, for the Doomsday argument to work, you must assume that every advanced alien civilization is going to eventually annihilate itself. This appears to push credibility to the breaking point. However, our best course of action appears to be to heed Stephen Hawking's warning.

You do not typically invite strangers into your house. Since you do not know them, you do not know their intentions. Why then should we invite strangers to our planet?

- **Extinction events**—Proponents of this argument suggest that eventually a natural event happens, like a giant asteroid hitting the Earth, destroying the majority of complex creatures. This is plausible. It happened to the dinosaurs 65 million years ago. It is a perilous universe. In addition to killer asteroids, a number of other extinction events are possible, like super volcanos and gamma-ray bursts. That Earth eventually will face an extinction event is a possibility. However, this time technology is in our corner. For example, we are watching for Earth-killer asteroids. These are asteroids that have some probability to strike the Earth, and are large enough to cause massive destruction. We are studying methods for diverting their course so they miss Earth. We are beginning to colonize space. Space stations have been in low Earth orbit since 1971. The latest manned space station, International Space Station (ISS), has been operational since 1998, and continuously occupied for over 11 years. In time, we will have space stations on the moon, Mars, and likely elsewhere. This makes wiping out all humankind by one extinction event less likely, since potentially large populations of humankind will be living on many celestial bodies, in addition to Earth. Therefore, if we can survive for approximately another century, we can likely render the probability of an extinction event destroying all humankind to be zero.

Now, we'll turn our attention to next key argument posited to resolve the Fermi paradox: the universe has intelligent life, but we see no evidence of it. Why would this be the case? Consider these reasons:

- **The immense universe theory**—Proponents of this theory argue that the scale of the universe is immense, and all intelligent life may be spread extremely distant from each other. For example, suppose Kepler 22b has intelligent life on it similar to our own. Even if, like us, they began to send radio transmissions a century ago, it will be another 500 years before we receive them, since Kepler 22b is 600 light years away. As mentioned earlier, by galactic standards, 600 light years is close. The Milky Way galaxy is 100,000 light years across. When the signal arrives, we may be extinct, or the Kepler 22b inhabitants may be extinct. Even if they are not extinct, our reply will take 600 years to reach them. The human species dates back about 200,000 years on our planet. The invention of radio is a little over a century old. In cosmic time, a century or even 200,000 years is a relatively small period. The universe is 13.7 billion years old. We may not have been around long enough or have been technologically capable long enough to detect advanced aliens or increase their interest in us. In addition, another hurdle has to be overcome. In Chapter 15, we discussed the universal currency of the universe is energy. On a comparative basis, sending a reply to any inhabitants of Kepler 22b is relatively inexpensive (uses less energy) versus sending a spacecraft to visit them. We likely have had the technology to colonize the moon for decades. Remember, Neil Armstrong put his footprint on the moon July 20, 1969. So, why didn't we colonize the moon? It comes down to economics, not technology. Colonizing the moon has no compelling economic reason. If we had found gold or diamonds plentiful on the moon, private enterprise would have wanted involvement. Private enterprise would be willing to underwrite a portion of the costs. If we had found a new energy source on the moon, the U.S. government would have made it a national priority. The fact is, all we found was moon dust and rocks. It is fascinating stuff, but not a sufficient reason to colonize the moon. This type of "economic" tradeoff is likely universal, and may explain why we do not see conclusive evidence of advanced aliens. They have no compelling reason to come to Earth or colonize it. They already have what they want, or can get what they want in their own galactic neighborhood, perhaps thousands of light years from us. The main point is that the immensity of the universe makes it difficult to find evidence of intelligent life.

- **Human communication technology is inadequate—**EarlierwritingaboutSETI,wediscussed the technical issues with receiving communications from advanced aliens,. To sum it up, the universe is immense. The resources of SETI are limited. Only a small portion of the universe can be listened to at a given time. The detection instruments may not have the necessary sensitivity or discernibility. In addition, advanced aliens may have moved beyond the need for radio technology. They may be radio silent and, for example, communicate with gamma-wave bursts or pulses. If we use ourselves as an example, consider the number of our telegrams versus telephone calls per year. The telegram is nearly obsolete. As we discussed above, perhaps aliens use gamma-ray pulses, and not conventional radio waves. In fact, we are outgrowing conventional radio and television broadcasts, as we move to fiber-optic cables, narrow-beam microwaves, and lasers. We became radio-visible on December 12, 1901, when Marconi sent radio signals from England to Canada. Our use of radio-visible signals peaked somewhere in the latter half of the Twentieth Century. It is currently over ten times less than what it once was. This is due to new, more energy-efficient technologies. In a relatively short while, we will almost be radio-silent. That means our radio visibility lasted about a century or so. Why would it be any different for advanced aliens?

Perhaps when they could broadcast, we did not know how to listen. Currently, we know how to listen, they no longer broadcast. Ironic, but possible.

- **Advanced aliens are part machine and part biology**—The idea here is that advanced aliens have extended their lifetime by replacing biological parts with bionic parts. They could have developed super computers that in turn developed even better super computers, and so on, and found ways to interface the super computers with their brains. In fact, they may be so intelligent that we appear to them like ants. We study ants, and even schoolchildren have or know about ant farms. Even though we understand that a form of intelligence allows ants to survive as a colony, we do not consider it worthwhile to make any effort to communicate with them. The prominent physicist, Michio Kaku, said it best in "Borrowed Time: Interview with Michio Kaku," Scientific American (November 23, 2003), "It's humbling to realise that the developmental gulf between a miniscule ant colony and our modern human civilisation is only a tiny fraction of the distance between a Type 0 and a Type III civilisation—a factor of 100 billion billion, in fact. Yet we have such a highly regarded view of ourselves, we believe a Type III civilisation would find us irresistible and would rush to make contact with us. The truth is, however, they may be as interested in communicating with humans as we are keen to communicate with ants." For clarification, a

Type 0 civilization represents the current technology capability of humans. A Type III civilization represents highly advanced aliens. When an intelligent life-form progresses significantly beyond their native intelligence and physical capabilities via extensive use of bionics, they are said to have experienced a technology singularity. Although this may read like science fiction, it is close to the direction we are heading. We are already making bionic replacements, for example, mechanical heart valves. As discussed in Chapter 7, based on Moore's law, desktop computers will become equivalent to human brains by 2029. (Moore's law states the number of transistors that can be placed inexpensively on an integrated circuit doubles approximately every two years.) Those machines will have the capability to improve themselves. By approximately 2050, super-computer intelligence will exceed human intelligence by orders of magnitude. Those machines might view us in much the same way we view seeing-eye dogs—as necessary aids to a blind person's quality of life. Therefore, if there are aliens a few centuries more advanced than us, their intelligence, knowledge, and capabilities could be orders of magnitude greater than our own.

- **Numerous other reasons**—This is the catchall category. For example, the advanced aliens are so different that they want nothing to do with us. Perhaps nothing about us matches up with anything about them, like sea-dwelling snails versus prairie bison.

Perhaps they do not speak at all, but use telepathy to communicate. Perhaps their communication machines are not radios, but a form of telepathy transmitters. Perhaps they purposely choose not to communicate, but listen. That is what we are doing. SETI listens. In a few cases have there been transmissions. The current official Earth policy is not to respond to any signal until we have consulted with numerous other nations of the Earth. Perhaps they consider Earth a zoo. Perhaps they are a peaceful race, and fear us. In other words, we may be like the junkyard dog. It could be that they have put no effort into weapons, and we are ahead in weapon technology. Therefore, while nuclear reactors power their homes and machines, they do not have any nuclear bombs. Perhaps they have made contact with a species like humankind in the past, and the outcome was terrible for all concerned.

My favorite one of all is: we have the evidence, but the government is suppressing it. It may be true, but it stretches credibility to the limit. Governments are generally not that adept at keeping secrets. Soviet spies stole the secret of the atom bomb as it was being developed during World War II. I understand that the United States government spends an enormous amount of money on "black" programs, which few people are cleared at a security level to see, along with the need to know. Therefore, it is possible, but would be the best-kept secret of all time.

Is there extraterrestrial intelligent life in the universe? We cannot definitively answer the question because definitive proof is not available. This chapter delineates most of the scientific highpoints known about the potential existence of extraterrestrial intelligent life. True, I did not cover some UFO data and other evidence. I am not casting dispersions on UFO data or any other evidence. It is not my area of expertise. My thrust was to address the question regarding the existence of extraterrestrial intelligent life from the standpoint of the physical and cosmological sciences. From that standpoint, given the information provided in this and previous chapters, we can make inferences regarding their existence. Based on all information presented, I infer that extraterrestrial intelligent life exists. You have the right to make your own inferences. I respect whatever conclusions you draw.

Next, we are going to investigate a mystery a little closer to home. It is not a light year away. It may be less than an inch away. In fact, it is playing out in front of us as you read these words. The mystery is best illustrated by asking a simple question, but the answer may not be simple. The answer could have profound consequences. The question is: what does the future hold for humankind? It is the subject of our next chapter.

"The future belongs to those who believe in the beauty of their dreams."

Eleanor Roosevelt (1884-1962)

CHAPTER 22

What does the future hold for humankind?

An old saying states, "The future is what we make of it." If ever that saying applied, it applies to the question: what does the future hold for humankind?

In reality, all we are able to see and measure in the universe will eventually end. To illustrate this, consider two points made in previous chapters.

1) **The expansion of the universe is accelerating**. Unless some unknown cosmic force causes this acceleration to slow down or even stop, eventually all that will be left of our universe will be the Milky Way galaxy. Every other galaxy will have moved beyond our cosmological horizon (beyond a point where we

can observe them due to the finite speed of light). All evidence of a Big Bang will be gone. The temperature of the universe will plummet to near absolute zero. The entropy (a measure of the disorder in a system) of the universe will reach a maximum, which I termed the "entropy apocalypse."

2) **Our sun is about five billion years old, and halfway through its life**. In another five billion years, the sun will die, and will destroy the Earth as it dies.

The above two cosmic events appear inevitable. They belong to a class of events termed "existential risks," which are any risks that could lead to the extinction of humankind. As bad as the two points above may seem, other existential risks exist that could cripple or destroy humanity, destroy the Earth, or even destroy the solar system in the blink of a cosmic eye.

Existential risks fall into two categories, manmade and natural. To understand this, consider the following examples.

Examples of Manmade Existential Risks

- **Global warming**—The Earth has experienced climate changes for as long as humankind has kept records. Alarmingly, scientists can only explain the recent pattern of warming by including the greenhouse gases, emitted by humans, into their predictive models. Global warming threatens the extinction of numerous species, in addition to humankind.

The Fourth Assessment Report (AR4) of the United Nations Intergovernmental Panel on Climate Change (IPCC) gave a sobering report in 2007. Two crucial points come from the report. First, "Anthropogenic warming could lead to some impacts that are abrupt or irreversible, depending upon the rate and magnitude of the climate change." Second, "There is medium confidence that approximately 20-30% of species assessed so far are likely to be at increased risk of extinction if increases in global average warming exceed 1.5-2.5 °C (relative to 1980-1999). As the global average temperature increase exceeds about 3.5 °C, model projections suggest significant extinctions (40-70% of species assessed) around the globe." While opinions differ among scientists regarding the validity of the points raised in the 2007 IPCC Fourth Assessment Report (AR4), it does raise a flag for concern.

The UN's Office for the Coordination of Humanitarian Affairs (OCHA) provided sobering statistics. According to OCHA, about 70% of disasters are currently climate related. Two decades ago, it was 50%.

As the Earth's average temperature increases, the polar ice caps will melt. This will cause sea levels to rise, and numerous coastal cities will be lost. The ratio of fresh water to salt water may change, endangering sea life, and potentially triggering another ice age. The Earth has been through 12 ice ages throughout its history. The last one occurred 10,000 years ago. All civilizations evolved after the last ice age. It is not clear that we can accurately predict when another ice age will occur. Before the intervention of humankind's greenhouse gasses, ice ages occurred about every 50,000 to 100,000 years. With humankind's greenhouse gas emissions, perhaps the next ice age will be hastened by unknown factors. We may not know when it will happen, but we believe it has the capability to destroy cavitation as we know it. Most of North America, Europe, and Asia could become uninhabitable.

- **Nuclear war**—For approximately the last 40 years, humankind has had the capability to exterminate itself. Few doubt that an all-out nuclear war would be devastating to humankind, killing millions in the nuclear explosions. Millions more would die of radiation poisoning. Uncountable millions more would die in a nuclear winter, caused by the debris thrown into the atmosphere, which would block the sunlight from reaching the Earth's surface. Estimates predict the nuclear winter could last as long as a millennium.
- **Bioterrorism**—The use of biological agents, as a tool for bioterrorism, dates back to Ancient Rome. The Roman soldiers would throw feces into faces of enemies. This crude but-effective method continued to be used by numerous armies until about the Fourteenth Century, when a better biological agent

became available—the bubonic plague. The use of the bubonic plague as a weapon wiped out large populations across Europe. In the Nineteenth Century, anthrax was discovered, proving to be an even better biological agent. Anthrax was better because it did not spread by person-to-person contact. To become infected required that the person or animal had to come into direct contact with anthrax spores. Progress continued in the Twentieth Century with the development of biological agents that killed animals, and did not affect humans. One such biological agent is the foot-and-mouth disease virus. The idea here is to create panic and economic ruin.

This is a thumbnail sketch intended to paint a picture of the potency of biological weapons. In fact, the potency of biological weapons to wipe out humankind was formally recognized by United States President Richard M. Nixon, who stated at a 1969 press conference, "Biological weapons have massive, unpredictable, and potentially uncontrollable consequences." He added, "They may produce global epidemics and impair the health of future generations." In 1972, President Nixon transmitted the Biological Weapons Convention to the U.S. Senate: "I am transmitting herewith, for the advice and consent of the Senate to ratification, the Convention on the Prohibition of the Development, Production, and Stockpiling of Bacteriological (Biological) and Toxin Weapons, and on their Destruction, opened for signature at Washington, London and Moscow on April 10, 1972. The text of this Convention is the result of some three years of intensive debate and negotiation at the Conference of the Committee on Disarmament at Geneva and at the United Nations. It provides that the Parties undertake not to develop, produce, stockpile, acquire or retain biological agents or toxins, of types and in quantities that have no justification for peaceful purposes, as well as weapons, equipment and means of delivery designed to use such agents or toxins for hostile purposes or in armed conflict."

While the 1972 convention is an enormous step forward for humankind, it has not deterred terrorists from using biological agents. For example, three bioterrorism attacks have occurred since its ratification. In 1984, politically motivated terrorists used Salmonella typhimurium bacteria to infect 751 people in the city of The Dalles, Oregon. Although this caused serious food poisoning, there were no fatalities. In 1993, terrorists attempted to attack the population of Tokyo using anthrax, but failed. The strain of anthrax they used was the vaccine strain of the bacterium. In 2001, the United States suffered anthrax attacks via postal letters laced with anthrax. At least five people died, and the terrorists evaded capture.

- **Technology singularity revolt**—The idea here is that technology improves itself beyond the limits of humankind's ability to control it, and the technology becomes adversarial.

One example is computers that become so intelligent they view us as a danger to their existence. Since computers have become a ubiquitous and indispensable part of advanced human civilization, a super computer revolt could potentially strike a devastating blow to civilizations throughout the world.

About 1970 a new form of technology began to emerge, nanotechnology. It had to do with building electro-mechanical structures on the atomic and molecular level. This is not science fiction. It is science fact. Nanotechnology is in various products, such as integrated circuit pressure sensors and printing heads for Inkjet printers. It has become a critical technology worldwide. The threat has to do with building nanotechnology devices that are self-replicating robots (nanobots). A typical example of this type of extinction event is the "grey goo" scenario. It goes something like this: someone builds a nanobot and programs it to build another nanobot, like itself, using whatever material it can find. Thus, the first nanobot builds another like itself. The two nanobots each build another. This soon gets out of control, and consumes all matter on the Earth. The Earth becomes infested with the grey goo of nanobots. Other nanotechnology threats, such as making nanobot weapons, are harmful to humans. This is the stuff of science fiction, where a human is infected with nanobots. Though it is still science fiction today, it will likely be science fact by 2050, based Moore's law, similar to our discussion in Chapter 7, regarding artificial intelligence.

Another possibility, similar to biological agents, is the creation of a super intelligent computer virus. I think of this as "techno-terrorism." This is a scenario: A super-intelligent computer virus is developed. It is unleashed into the computer population of the world, which includes one billion personal computers as of 2012, not counting smartphones. It spreads undetected, computer to computer, for years. Then, at a predetermined critical infection level, it becomes active, and renders the computers we use to sustain civilization unusable. Computers all over the Earth will be rendered useless. Civilizations become destabilized, and crumble.

Stephen Hawking weighed in on this one, stating, "I think computer viruses should count as life. I think it says something about human nature that the only form of life we have created so far is purely destructive. We've created life in our own image." I've quoted Dr. Hawking to illustrate that we face a real and present danger. I am not quite as pessimistic in my view as Dr. Hawking. I ascribe to Dr. King's view of the future, discussed below.

Examples of Natural Existential Risks

- **Super volcanos**—A super volcano is one capable of ejecting 1,000 cubic kilometers (240 cubic miles) of magma and ash into the atmosphere. This is thousands of times more material than the Mount St. Helens Volcano eruption in 1980. The ejection of this much material into the atmosphere has the potential to cause a Volcanic Winter, where the densely polluted atmosphere does not permit sufficient sunlight to reach the Earth's surface. To date, six locations potentially harbor a super volcano. The best-known location is Yellowstone National Park. The last time Yellowstone erupted was 640,000 years ago, when its magma and ash covered most of the United States west of the Mississippi, as well as portions of northern Mexico.
- **Asteroids**—65 five million years ago, the most successful animal species on the Earth, the dinosaurs, was wiped out by a large asteroid, approximately six miles in diameter. Asteroids continually collide with Earth. Most are small, similar to the size of dust that accumulates on furniture. Most burn up in our atmosphere. When an asteroid is large enough to penetrate the Earth's atmosphere, without burning up, and it impacts the Earth, it is typically termed a "meteorite." An asteroid would need to be at least .6 of a mile in diameter to significantly disrupt civilization. Fortunately, the probability of that size asteroid striking Earth is small. Current estimates are

that an asteroid of .6 miles in diameter will strike the Earth every 500,000 years. Large asteroids, 2-3 miles in diameter, will strike every 10,000,000 years. You may be wondering how we know these statistics, and calculate the probabilities. It is based on measuring the crater diameters on the moon. From moon-crater data, we know there is an inverse relationship between the size of the asteroid, and its probability to strike the Earth. In other words, the larger the asteroid, the less probable it is to strike Earth. However, numerous people have seen a "falling star" in the night sky. These are small asteroids, which burn up in the atmosphere. In addition to asteroids, other celestial bodies can have a devastating effect if they strike Earth, such as comets.

- **Collisions of our galaxy (Milky Way) with the Andromeda Galaxy**—This may at first sound like science fiction. However, based on astronomical observations, the Andromeda Galaxy is on a trajectory course to collide with the Milky Way galaxy. The two galaxies will collide in about 3 billion years. Collision of galaxies is not a hypothetical event. Astronomers have observed and photographed this type of collision. Astronomers claim galaxy collisions are common and that, at one time, every galaxy interacted with it neighboring galaxies.

Obviously, I have provided a handful of existential events that could spell doom for humankind. There are more. The fact is we live in perilous times. We have always lived in perilous times, but the game changer is our own ability to annihilate ourselves. We have had this ability for about half a century, assuming the ability to arm intercontinental ballistic missiles with nuclear weapons is the demarcation line. What do we do about existential events? To a high probability, the future is in our hands. Three courses of action, taken as a whole, are essential to the survival of our species, namely:

1) End all wars—All wars divert resources that could benefit humankind. We must stop fighting with each other, on every level. It is necessary for our survival. Think back. Has any war, in human history, settled anything? Did World War I settle things? That was supposed to be the war to end all wars. It failed. Along came World War II. Did it settle anything? Almost everyone the United States was fighting in World War II is currently an ally of the United States. War is futile, wasteful, and potentially spells doom for humankind. I recognize that people around the world are working toward ending all wars. The United Nations is an excellent example. To a limited extent, they have been successful. We have avoided a third world war. Today's wars are relatively limited in scope and use of weapons. However, we are still a long way from being a peaceful planet. Since we have the capability to make humankind extinct, even limited wars pose substantial risks to humankind. Even a "limited nuclear war" may ignite an all-out nuclear

war. One bioterrorist act may spread globally. In the Cold War (1940s-1990s) between the United States and the Soviet Union, each country had the ability to initiate an all-out nuclear war. However, that never happened because of mutually assured destruction (MAD). In other words, there would be no winners. That is where we are today. Getting this message out in an understandable way to all humankind may go a long way in charting a new course to peace and a more hopeful future.

2) Colonize space—Stephen Hawking made a startling prediction, "I don't think the human race will survive the next thousand years, unless we spread into space." To Dr. Hawking, the threat is a real and present danger. If you agree, we must continue our efforts to spread our species throughout the galaxy, and eventually to other galaxies. We can start by establishing civilizations within our own solar system. As technology permits, continue our expansion to nearby solar systems, and some day, have entire Earth-like planets within our galaxy colonized. Technology will eventually enable us to make this a reality. Perhaps we will even learn how to travel to a parallel world. Today's science fiction is tomorrow's science fact.

Another option is to become nomads. This is the vision Stephen Hawking painted of the advanced aliens, when he warned humankind against contacting them. The idea is that humankind would build city-size or larger spaceships, and travel the galaxy, stopping as needed to replenish supplies, make repairs, and undertake numerous other activities, perhaps even including leisure. Is this too far out to be believable? *New Internationalist Magazine*, Issue 266, reported approximately 30-40 million nomads on Earth in 1995. However, most experts believe that number is dropping. Unfortunately, there does not appear to be a definitive up-to-date headcount. This should not surprise us since by their cultural nature, nomads would be hard to count. Why do relatively large groups of people become nomads? The answer is one word: survival. Nomadic people's ability to move locations favors their survival. In addition, typically their tradition is to form clans or tribes, with hierarchies and rules. This sounds a lot like the way early humans inhabited the Earth. Why would it be so far out to believe that we might choose to inhabit the galaxy as nomads, prior to colonizing the right location? This would be especially true if the journey to the right location would take hundreds to thousands of years.

3) Mitigate existential events—Use technology to divert large asteroids on a trajectory to strike the

Earth. Find ways to harness the energy of potential super volcanos, making them work for us rather than against us. A lot of work is being done in the area of mitigating existential events. Numerous organizations have formed to address specific existential events. These include NASA's Near Earth Object program, the Singularity Institute for Artificial Intelligence, the Center for Responsible Nanotechnology, and the Svalbard Global Seed Vault, to name a few. We are moving forward, but the strategic and ethical ground is still cloudy. What should we do first, and why? Nick Bostrom, a Swedish philosopher at Oxford University, and one of the best-known authorities on existential risks, gave us a useful rule-of-thumb for moral action. In his paper, Existential Risks (Journal of Evolution and Technology, Vol. 9, No. 1 [2002]), Professor Bostrom states, "Previous sections have argued that the combined probability of the existential risks is very substantial. Although there is still a fairly broad range of differing estimates that responsible thinkers could make, it is nonetheless arguable that because the negative utility of an existential disaster is so enormous, the objective of reducing existential risks should be a dominant consideration when acting out of concern for humankind as a whole. It may be useful to adopt the following rule of thumb for moral action; we can call it Maxipok: Maximize the probability of

> an okay outcome, where an 'okay outcome' is any outcome that avoids existential disaster."
>
> This sounds simple, doesn't it? All we have to do is get all nations of the world and all groups within those nations to maximize the probability of an okay outcome. However, getting the world aligned on the Maxipok is the equivalent of ending all wars. Ending all wars is a simple concept. It is logical. It is ethical. It is required for humankind's survival. Yet, it is hard to get the world aligned, even when it is in our own best interest.

Obviously, humankind faces an unprecedented number of existential risks capable of rendering our species extinct. Several existential risks are manmade, such as nuclear war and bioterrorism. Others are natural or cosmic-born, such as super volcanos erupting, and large asteroids striking the Earth. For the first time in the history of our species, we have the capability to mitigate some, and perhaps, in time, all existential risks. However, it is a choice we as a species will have to consciously make. As Martin Luther King, Jr. put it, "I refuse to accept the view that mankind is so tragically bound to the starless midnight of racism and war that the bright daybreak of peace and brotherhood can never become a reality... I believe that unarmed truth and unconditional love will have the final word."

As I wrote in the first paragraph, the future is what we make of it. I believe the true spirit of humankind is to seek peace, and to care for one another as Dr. King so eloquently stated.

Next, we'll ask a crucial, but controversial question: how does God fit into the equation? It is the subject of the last chapter.

"It is as impossible for man to demonstrate the existence of God as it would be for even Sherlock Holmes to demonstrate the existence of Arthur Conan Doyle."

Frederick Buechner (American Author, b.1926)

CHAPTER 23

How Does God Fit into the Equation?

No scientific equation allows us to prove or disprove that God exists. If there were such an equation, the "God versus Science" debate would be over.

- Are there philosophical proofs of God's existence? Yes! Are they refutable? Yes, numerous philosophers and scientists refute them.
- Are there "scientific" proofs that God exists? Yes! The scientific proofs are not equations. They generally argue philosophically using one or more laws of physics as their pillars. Are they refutable? Yes,

numerous scientists and philosophers refute them. It appears that equally intelligent, knowledgeable and sincere people are on both sides of this debate.

God versus science is a debate that has been going on for centuries. It started at the very birth of science. I believe that you will find this chapter insightful. However, it will not settle the debate. It will frame the debate, and ask you to consider a serious question that is central to the debate: is it possible to prove or disprove the existence of God using the natural sciences? Addressing this question is the key thrust of this chapter. The answer you arrive at will determine if the debate is necessary. It will assist you in deciding if you want to engage in the debate. Therefore, we will examine the evidence, not to prove or disprove God exists, but to address whether it is possible to prove or disprove God exists. To address this question, we will have to start with fundamentals.

First, we need to start with a definition of God. God means different things to different people, cultures, and civilizations. We likely could fill hundreds of pages, and still not cover all the potential definitions. Therefore, in the interest of brevity and focus, we will not concentrate on the beliefs of any one religion, but rather on the general themes regarding the nature of God that weaves through numerous religions. For this reason, it would be better to not use the word God, since it is often associated with monotheism (one God), but rather the word "deity," which encompasses polytheism (multiple Gods). The question becomes, what is the nature of a deity?

Five key attributes of a deity are found in numerous religions throughout the world:

1) It is an eternal, divinely simple (no parts), supernatural being (omnipreternatural).
2) It knows all (omniscience).
3) Its power is unlimited (omnipotence).
4) It is everywhere (omnipresence).
5) It is all good (omnibenevolence).

If we attempt to go beyond the five attributes described above, we inevitably get into specific religions. However, the five attributes are sufficient to make one pivotal point, namely that a deity is supernatural. What would this imply? It implies that a deity is beyond the realm of nature (physical reality). Therefore, any experiments we do in the physical world will give us physical data, not supernatural data. This is a critical point; it implies that scientific proof of the existence of a supernatural being is impossible. The reverse is true. It implies it is impossible to disprove the existence of a supernatural being. We could stop here because the objective was to answer the question: is it possible to prove or disprove the existence of God using the natural sciences? If you assume God is supernatural, item 1, using the natural sciences to prove or disprove God's existence would appear futile. In the interest of more fully addressing the question, we will examine the attributes of a deity a little closer.

What about attributes 2 through 4 (omniscience, omnipotence, omnipresence)? Given sufficient time, technology would give humankind or advanced aliens the capabilities ascribed to items 2 through 4. For example, an advanced alien with a billion-year technology jump on us would appear like a deity to us. This is Clarke's Third Law (Arthur Charles Clarke, 1917-2008, British author, inventor and futurist), namely, "Any sufficiently advanced technology is indistinguishable from magic." Prominent physicist, Michio Kaku, made a startling prediction in his book, Physics of the Future, "By 2100, our destiny is to become like the gods we once worshipped and feared. But our tools will not be magic wands and potions but the science of computers, nanotechnology, artificial intelligence, biotechnology, and most of all, the quantum theory."

What about attribute 5 (omnibenevolence—it is all good)? Before we can address attribute 5, we need to understand what the word "good" means in this context. After much thought and research, I suggest that the word "good" has no meaning in the context of a deity. In fact, I suggest that the word "benevolence," which can mean good, kind, charitable, and so on, has no meaning in the context of deity. Please consider this example: Is a cat "good" when it catches a mouse for food? To my mind, it is neither good nor evil, since the cat is acting in accordance with its nature, and within the laws of nature. It must catch mice to survive. However, a deity by definition has no nature, no set of laws, and no boundaries of any kind. It is. Whatever it does, is. If a deity exists, we exist because the deity enables it. The deity defines our perceptions of good and evil.

Where does this leave us? It leaves a lot of scientists and

others with ruffled feathers. Let me frame the debate. From a historical broad-brush perspective, comes two main camps:

1) Evolutionism—asserts we are here because of evolution, not divine intervention. In a modern context, the evolution is viewed to have started with the Big Bang itself. This school has also been termed "Darwinian evolution" and "scientism."
2) Creationism—asserts we are here because of divine intervention, referred to as "intelligent design." In a modern context, the Big Bang was "God's" chosen method to create our universe, which ultimately resulted in our existence

Is this a quiet, low-key debate? Does it confine itself to discussions at cocktail parties or polite conversation among academies? Not by a light year. A recent argument put forward by the creationists to "prove" the existence of God is called "Intelligent Design." Intelligent Design argues that Intelligent Cause alone can explain the universe. In other words, existence is not the evolution of chance, but the result of intelligent design. Creationists developed this theory to get around the Supreme Court ruling that barred the teaching of "Creation Science" in public schools, based on the ruling it breached the separation of church and state. In effect, Creationists put forward Intelligent Design as a scientific theory. However, scientific theories must be testable. It is okay to imagine and conjecture in science, but if you want to forward a theory, it must be testable. For this reason, mainstream science rejects Intelligent Design because it lacks experimental verification, and by its nature is untestable. It failed to accomplish its objective, namely to get a version of "Creationism" into public school curriculums. In 2005, in the Kitzmiller versus Dover Area School District trial, U.S. District Judge John E. Jones III ruled that Intelligent Design is not science, that it "cannot uncouple itself from its creationist, and thus religious, antecedents." Once again, it was removed from school curriculums on the basis that it violated the First Amendment to the U.S. Constitution, which specifically prohibits Congress from preferring one religion to another.

As things were quieting down a bit, Stephen Hawking, the world's most famous scientist, made a startling statement on September 2, 2010, one week prior to the release of his new book, *The Grand Design*. He declared the "Almighty" irrelevant. Dr. Hawking believes that M-theory may hold the ultimate key to understanding everything, even the birth of the universe. Therefore, the need for religion becomes unnecessary. Of course, critics ask where M-theory came from. This is surprising since Dr. Hawking is on record saying, "Even if there is only one possible unified theory, it is just a set of rules and equations. What is it that breathes fire into the equations and makes a universe for them to describe?" To my mind, this is the right question.

Dr. Hawking is just one scientist, albeit highly famous. In general, what do scientists believe? Numerous studies, regarding scientists in the United States, indicate about a third are atheists, a third agnostic, and a third believe in God or a higher power. Similar studies of the general population suggest that three-fourths of the population believes in God or a higher power. (Survey 2005-2007 by Elaine Howard Ecklund of University at Buffalo, The State University of New York)

What does this mean? A majority in the scientific community appears to be rebelling. According to *Time* magazine, "God vs. Science," Dan Cray/Los Angeles, Nov. 05, 2006, the rebellion is due to, "what one major researcher calls 'unprecedented outrage' at perceived insults to research and rationality, ranging from the alleged influence of the Christian right on Bush Administration science policy to the fanatic faith of the 9/11 terrorists to intelligent design's ongoing claims." In addition, the majority scientists no longer look to religion for answers, but to their science.

The elegance and orderliness of scientific theories and mathematics becomes seductive and, in effect, replaces a need for a higher deity. However, this is not to say there is any unified conspiracy on the part of the scientific community to replace religion with science. In fact, without intention, science and religious ethics appear to have much in common. Einstein wrote in "Essays in Physics" (1950), "However, all scientific statements and laws have one characteristic in common: they are "true or false" (adequate or inadequate). Roughly speaking, our reaction to them is "yes" or "no." The scientific way of thinking has a further characteristic. The concepts which it uses to build up its coherent systems are not expressing emotions. For the scientist, there is only "being," but no wishing, no valuing, no good, no evil; no goal. As long as we remain within the realm of science proper, we can never meet with a sentence of the type: "Thou shalt not lie." There is something like a Puritan's restraint in the scientist who seeks truth: he keeps away from everything voluntaristic or emotional."

However, regardless of the inherent ethics, shared by science and religion, one thing that stands in the center of this passionate debate is the existence of miracles. For something to be a miracle, it must be outside the natural laws of science. In effect, natural law is suspended, and a miracle happens. A majority of scientists have difficulty believing this. Einstein summed this up in the following statement, "Development of Western science is based on two great achievements: the invention of the formal logical system (in Euclidean geometry) by the Greek philosophers, and the discovery of the possibility to find out causal relationships by systematic experiment (during the Renaissance)." To illustrate the difficulty of suspending natural laws, consider this example. If I told you apples fall up instead of down, would you believe me? Probably not. You probably would not even argue with me. My guess is that you would likely be dismissive, and ignore me. Yet, at the heart of various religions is the belief in miracles.

Is it possible to suspend natural laws? I suspect most scientists would answer a resounding "No!" However, in Chapter 2, we asserted the formation of the infinitely dense energy point, which led to the Big Bang, was due to a quantum fluctuation, similar to the formation of virtual particles. In effect, this appeared to suspend the conservation of energy law. However, it does not. I argued that the quantum fluctuation resulted in an infinitely energy-dense matter-antimatter particle pair, and not a single infinitely dense particle. I did this for a reason. It maintains the conservation of energy. We are not creating energy since the energy of the system remains unchanged. The matter-antimatter particle pair cancels each other out since they are the equal and opposite of each other. In addition, most scientists ascribe the formation of virtual particles to be in accordance with the Heisenberg Uncertainty Principle, a law of quantum mechanics. Dirac postulated that a vacuum was filled with matter-antimatter pairs (electrons and positrons), which he called the Dirac sea. In effect, the energy was already there, but neutral and undetectable. Does science thoroughly understand this? Not really, We may not entirely understand all the laws of physics, and their relationship to each other. However, nature follows all physical laws, even the ones we do not understand. Consider this example: Advanced aliens may have a science that appears to suspend natural laws. Perhaps they know how to create "worm holes" and travel vast distances, faster than the speed of light. To our observations, they may be violating another pillar of modern physics, namely the speed of light in a vacuum is the upper limit of velocity in the universe. However, simply because we do not understand their science does not mean that they have suspended natural law. They simply have learned secrets about nature we have not discovered. They know how to harness more energy than we do, which allows them to apparently violate nature laws and create miracles. This may make them appear god-like. However, they are not the god (deity) that most scientists find problematic. I have not seen an actual survey on this, but I would be willing to hazard a guess that more scientists believe in advanced aliens than believe in a deity. The existence of miracles and the suspension of natural laws, rubs scientists the wrong way. That is one of the significant dilemmas that today's scientists face when they try to reconcile their science with their faith.

This debate is essentially unresolvable. In the "evolutionary camp," you have scientists and philosophers that believe they can explain our existence without "creation," using the natural sciences and reason. In the "creation camp," you have philosophers and scientists that refute creationism, based on plausible logic and reason. We have gridlock, evolutionists versus creationists. Since this has been going on for centuries, I do not think advances in science or interpretations of scripture are going to end the gridlock.

What does this boil down to if we take the personalities at the center of the current debates out of the picture? The nature of being "God" implies a supernatural being. Science deals with natural phenomena. Logically, it appears irrational to believe that science, which attempts to understand, model, and predict natural phenomena, can be used to investigate supernatural phenomena. Obviously, if the existence of God were provable, religious leaders would not ask for faith. There have been proofs of God dating centuries back. For example, Thomas Aquinas in 1270 published *Summa Theologia*, which delineated five proofs of God's existence. Obviously, it was compelling for countless people, but not others. From that perspective, it is up to the individual. It is a choice, to believe or not to believe. Conversely, science does not require belief as the final step in the process. Belief plays a role in science, especially as new theories surface, but ultimately scientists seek experimental verification.

Personally, I have chosen not to enter the debate, reasoning I have no way to resolve the issue on either side. Scientists recognize that ultimately we are trying to explain, predict, and model physical reality. Therefore, ultimately scientific theories are subject to experimental verification. This does not require unconditional faith. To believe in a deity requires unconditional faith, which is born from belief. In the United States, you are free to hold any religious beliefs. You are free to believe or not believe in a deity. You are free to worship, as you deem appropriate. If your beliefs differ from others, it is your right, and the United States Constitution protects your right. Wherever you stand on the question of a deity's existence, I respect your right to hold whatever beliefs you embrace, and to worship as you choose.

Einstein gave us one valuable insight that might help calm the debate, namely, "All religions, arts and sciences are branches of the same tree. All these aspirations are directed toward ennobling man's life, lifting it from the sphere of mere physical existence and leading the individual towards freedom." This leaves me with a question: do we need a debate?

We are coming to an end of this book, and I would like to leave you with several closing thoughts. These will be the significant "take-aways" from the last twenty chapters. However, please understand this is one book out of a vast sea of books on scientific mysteries. I hope it has provided insight, and increased your curiosity. However, it is far from the last word on the subject of scientific mysteries. I referenced numerous excellent scientists and philosophers that have insights you may wish to delve into further. We will distill the main points of this work in the next section, "Closing Thoughts."

Closing Thoughts

What is new in this book? What major new theories and concepts surfaced for consideration? This is what I intend to cover in my closing thoughts. My objective is to reveal the six scientific "nuggets" worthy of consideration, with the objective of sparking a dialogue and further research:

1) **The Big Bang Duality theory**

   Rationale of importance:

   The Big Bang Duality theory explains the origin of the Big Bang. It postulates the Big Bang is due to the collision of infinitely energy-dense matter-antimatter particles in the Bulk (super-universe). In addition, it suggests that the physical laws of our universe originate in the Bulk. Lastly, the Big Bang Duality theory explains the absence of antimatter in our universe, without requiring a violation of the fundamental symmetry of physical laws.

   Discussion:

   It is reasonable to consider that a quantum fluctuation in the Bulk resulted in an infinitely energy-dense particle-antiparticle pair, not a single infinitely energy-dense particle. This equates to an energy neutral system, and aligns with the conservation-of-energy law.

If the quantum fluctuation theory is correct, it makes a strong case that the scientific laws of our universe are the scientific laws of the Bulk. This implies the physical laws of the universe pre-date the Big Bang, and that if there were other universes created via quantum fluctuations, they too would follow the laws of the Bulk.

Lastly, by postulating a spontaneous creation of infinitely energy-dense matter-antimatter particle pairs that collide in the Bulk to create what is commonly referred to as the Big Bang, we are able to explain the absence of antimatter in our universe. In effect, it was consumed during the initial matter-antimatter particle collision and the subsequent interactions. This model, unlike other models of the Big Bang, does not require a violation of the fundamental symmetry of physical laws.

2) Minimum Energy Principle

Rationale of importance:

The Minimum Energy Principle states: *Energy in any form seeks stability at the lowest energy state possible and will not transition to a new state unless acted on by another energy source.* This implies the Big Bang went "bang" at the instant it came to exist.

Discussion:

The Minimum Energy Principle is a generalized statement of similar laws in the physical sciences. In its current formulation, it is independent of the scientific context.

3) **Consider dark matter a form of energy, not a particle.**

Rationale of importance:

This provides a new thrust for research, and explains why the Standard Model of particle physics does not predict the dark matter particle—WIMP (weakly interactive massive particle). In addition, it explains why efforts to detect it have been unsuccessful.

Discussion:

The existence of dark matter is not in dispute. However, serious efforts to prove that dark matter is a particle—WIMP (weakly interactive massive particle) —have been unsuccessful. In fact, The Standard Model of particle physics does not predict a WIMP particle. The Standard Model of particle physics, refined to its current formulation in the mid-1970s, is one of science's greatest theories. If the Standard Model does not predict a WIMP particle, it raises serious doubt about the particle's existence. All experiments to detect the WIMP particle have, to date, been unsuccessful. Major effort has been put forth by Stanford University, University of Minnesota, Fermilab, and others to detect the WIMP particle. Millions of dollars have been spent over last decade to find the WIMP particle. Despite all effort and funding, there has been no definitive evidence of its existence. This appears to beg expanding our research scope. One approach suggested is that science attempt to model dark matter using M-theory.

4) The Existence Equation Conjecture

Rationale of importance:

The Existence Equation Conjecture is, arguably, the most important theory put forward in this book. It relates time, existence, and energy. It explains the physical process related to time dilation. It rests on three pillars:

1) The fourth dimension, although a spatial coordinate, is associated with existence in time.
2) Movement in the fourth dimension (existence) requires enormous negative energy as suggested by the Existence Equation Conjecture ($KE_{X4} = -.3mc^2$).
3) When we add kinetic energy or gravitational energy to a particle, we reduce the amount of negative energy it requires to exist and, thus, increase its existence.

Discussion:

This equation is dimensionally correct, meaning it can be expressed in units of energy, which is an important test in physics. The equation is highly unusual. First, the kinetic energy is negative. Second, the amount of negative kinetic energy suggested by the equation, even for a small object like an apple, is enormous. The energy, for even a small object, is about equivalent to a nuclear weapon, but negative in value. This led me to postulate that the source of energy to fuel the Existence Equation Conjecture is dark energy. Modern science believes dark energy is a negative (vacuum) form of energy causing space to expand. From the Existence Equation Conjecture, we know existence requires negative energy to fuel existence. Comparing the Existence Equation Conjecture's need for negative energy seems to suggest existence may be syphoning its required negative energy from the universe. This implies that existence and dark energy may be related.

In summary, we have a more complete picture of time's nature, namely:

1) Time is related to change (numerical orders of physical events)
2) Time is related to energy via its relationship to change, since change requires energy

3) Time is related to existence, and existence requires negative energy per the Existence Equation Conjecture

4) The energy to fuel time (existence) may be being acquired from the universe (dark energy), causing the universe to expand (via the negative pressure we describe as dark energy). This aligns conceptually with the form of the equation, and the accelerated change in the universe.

5) The enormousness changes in entropy (disorder) in the universe may be the price we pay for time. Since entropy increases with change, and time is a measure of change, there may be a time-entropy relationship.

The derivation and experimental verification of the Existence Equation Conjecture can be found in Appendices I and II.

5) The Quantum Universe theory

Rationale of importance:

This theory postulates that all reality, including space, consists of quantized energy.

Discussion:

The majority of experimental and theoretical data argues that the macro world, the universe in which we live, is the sum of all matter and energy quanta from the micro world (quantum level). Recent experiments demonstrate that the micro level and quantum level can influence each other, even to the point they become quantum entangled. In addition, space itself appears quantized, considering the Dirac sea, the particle theory of gravity, and the irreducible Planck length. This allows us conceptually to describe the universe as a Quantum Universe.

6) **The existence of God (deity) is not scientifically provable**

Rationale of importance:

This debate, God versus Science, is centuries old. It revolves around the question: can science prove or disprove God (deity) exists? The effects of such a proof would be profound.

Discussion:

This debate is essentially unresolvable. The nature of being "God" implies a supernatural being. Science deals with natural phenomena. Logically, it appears irrational to believe that science, which attempts to understand, model, and predict natural phenomena, is extendable to investigate supernatural phenomena. Obviously, if the existence of God were provable, religious leaders would not ask for faith. It is a choice, to believe or not to believe. Conversely, science does not require belief as the final step in the process. Belief plays a role in science, especially as new theories surface, but ultimately scientists seek experimental verification.

A Few Words from the Author:

First, I hope you enjoyed the book as much as I enjoyed writing it. Perhaps I have sparked the curiosity of some readers to the point that they will consider a career in science.

I believe the universe is wondrous. I do not think we will ever know all there is to be known. As I stated early on, important scientific discoveries typically lead to profound scientific mysteries (Del Monte Paradox). Yet, we humans are a curious species and, in the long term, I sense our curiosity will work in our favor.

A truly enormous question holds the key to all scientific mysteries: will we ever get a theory of everything? Scientists of the stature of Stephen Hawking believe M-theory, the theory that ties all string theories together, may serve as the theory of everything. However, Dr. Hawking has run hot and cold on M-theory as the theory of everything. He is on record pondering if a theory of everything even exists.

If a theory of everything does exist, and explains all natural phenomena, I do not think it or any physical theory will prove or disprove the existence of God. If God exists, it is in the supernatural realm. Physical sciences deal with the natural realm. I recognize this logic is likely not going to end the debate, but I ask that you consider it before choosing to enter the debate.

From my research, I have come to believe that all reality is a form of energy, which inspired me to postulate that all reality is quantized. This led to the assertion of the Quantum Universe theory. As Einstein put it, "Reality is merely an illusion, albeit a very persistent one." Does energy create the illusion? Michael Faraday, in 1845, probably said it best, "I have long held an opinion, almost amounting to conviction, in common I believe with many other lovers of natural knowledge, that the various forms under which the forces of matter are made manifest have one common origin; or, in other words, are so directly related and mutually dependent, that they are convertible, as it were, one into another, and possess equivalents of power in their action." Although, I feel a cogent argument was made regarding the Quantum Universe theory, I recognize elements within the theory still require experimental verification. I ask that you consider it, and draw your own conclusions.

As this book ends, I want to leave you with a quote from Elizabeth Barret Browning, "Gaze up at the stars knowing that I see the same sky and wish the same sweet dreams." The sweet dreams I wish for you are peace and love.

Glossary

Absolute zero – The theoretical temperature at which all motion stops, even at the atomic level, and the substance has no heat energy. According to the laws of physics, it is not possible for a substance to reach absolute zero.

Accelerating expansion of the universe – It is a scientific fact that the expansion of the universe is accelerating. The farther away a galaxy is from us, the faster it is accelerating away from us. In fact, the acceleration of the farthest-away galaxies appears to exceed the speed of light. Modern science believes that it is space that is expanding faster than the speed of light, which makes it appear that the galaxies themselves are accelerating faster than the speed of light. No law of physics prevents space from expanding faster than the speed of light.

Acceleration – The change in velocity (speed) of an object over a specific time.

Advanced aliens – This refers to intelligent extraterrestrial life that is scientifically knowledgeable and able to apply technology, such as radio broadcasting.

Anthropic principle – This principle asserts the universe is the way it is because if it were different, we would not exist.

Anthropogenic warming – This is global warming caused or influenced by humans.

Antimatter – The mirror image of matter. In matter, the electrons in atoms are negative, and the nucleus is positive. In antimatter, the electrons in atoms are positive, and the nucleus is negative.

Baryogenesis theories – Refers to a class of theories that assert the existence of a process that creates an asymmetry between matter and antimatter in the universe. Baryogenesis theories are postulated to explain the absence of antimatter in the universe. These theories are unproven and considered speculative.

Big Bang duality – This theory postulates that the Big Bang did not originate as a singularity (one highly dense energy particle), but a duality (one highly dense energy particle and one highly dense energy antiparticle). The expansion of energy (commonly referred to as the Big Bang), according to the Big Bang Duality theory, was the result of a matter-antimatter particle collision in the Bulk.

Big Bang theory – The theory that the universe originated from a highly dense energy state that expanded to form all that we observe as reality.

Big Crunch – This theory held that gravity would eventually overcome the expansion of the universe, and force all reality into a highly dense energy point.

Black hole – Refers to a region of space whose gravitation is so intense that not even light can escape, thus the name "black hole."

Bubble Universes – Refers to entire universes that theoretically exist in the Bulk. The bubble universes may or may not resemble our own.

Bulk – A super-universe capable of holding countless universes. In theory, it contains our own universe, as well as other universes.

Casimir-Polder force – The attractive force between two parallel metal plates placed extremely close together (approximately a molecular distance) in a vacuum. Science believes the "attraction" is due to a reduction in virtual particle formation between the plates. This, in effect, results in more virtual particles outside the plates whose pressure pushes them together.

Celsius – A temperature scale, at which water freezes at zero degrees (0° C) and boils at one hundred degrees (100° C).

Chaotic inflation theory – This theory asserts that the universe does not inflate uniformly, but may accelerate in regions of space devoid of matter and radiation. The non-uniform inflation causes portions of the universe to separate from the existing universe. These portions form mini-universes ("bubble" universes). The word "bubble" is used to describe this concept because when bubbles form, occasionally a smaller bubble will form on a larger bubble, and then separate from it.

Chronology protection conjecture – This is a conjecture by physicist Stephen Hawking, published by him in a 1992 paper, which asserts the laws of physics prevent time travel on all but submicroscopic scales. This conjecture has been an object of debate, and has substantial supporters and adversaries. We have no consensus that it is a scientific fact.

Classical mechanics – Refers to Newton's three laws of motion, enunciated by Newton in the Seventeenth Century. It is widely used today by almost everyone since it provides excellent agreement with phenomena in our everyday experiences (the macro-world). For example, if you play billiards, Ping-Pong, or even marbles, you are intuitively using Newton's laws of motion. Newtonian mechanics and Newton's laws of motion are synonymous to Classical Mechanics.

Closed time-like curves – In mathematical physics, a closed time-like curve is a solution to the equations of general relativity that demonstrates a particle's world line closes on itself (returns to the starting point). Numerous theoretical physicists interpret this to imply time travel to the past is possible.

Conservation of energy – Arguably the most sacred law in physics, namely that energy cannot be created or destroyed, only transformed from one form to another.

Conservation of mass – This law is similar to the conservation of energy, namely mass cannot be created or destroyed, only transformed from one form to another, including the transformation to energy.

Cosmological constant – An arbitrary constant, originally proposed by Albert Einstein, to force his equations in general relativity to predict a static universe. In 1929, when Edwin Hubble discovered the universe was expanding, Einstein proclaimed the cosmological constant his "greatest blunder." Today, physicists are using the cosmological constant, along with mathematical manipulation of general relativity, to model the accelerated expansion of the universe.

Cosmological horizon – The observable universe. Its basis is the time light has had to travel to us from the edge of the observable universe, in accordance with the age of the universe (13.7 billion years old).

Dark energy – A theoretical force postulated to be responsible for the accelerated expansion of the universe. Scientists describe it as a "vacuum" force, while others describe it as a "negative" force. In reality, no scientific consensus exists regarding the nature of dark energy, including its existence.

Dark matter – This is mass in galaxies, galaxy clusters, and between galaxy clusters that cannot be observed directly, but is postulated to exist to explain why the farthest-away stars from the center of the galaxy rotate at approximately the same rate as the stars nearer the center of the galaxy. From the laws of physics, the farthest-away stars should be rotating much slower than those nearer the center. Therefore, scientists have postulated that more mass, namely dark matter, is in the galaxy than we observe since it did not emit or reflect light. The existence of dark matter has been confirmed by gravitational lensing (gravity bending light) observations. Scientific calculations indicate that dark matter may account for about 90% of the total matter of the universe. Scientific speculation asserts that a particle is associated with dark matter, namely the WIMP particle (Weakly Interacting Massive Particle). To date, there is no conclusive evidence that the WIMP particle exists.

Del Monte Paradox – An observation made by Louis Del Monte and introduced in this book, namely: Each significant scientific discovery results in at least one profound scientific mystery.

Dilated – Typically used as a verb in this book, and refers to one clock traveling at the speed of light running slower than another clock at rest.

Dirac sea – A theory postulated in 1930 by Paul Dirac (British physicist) that empty space (a vacuum) consists of a sea of virtual electron-positron pairs. This eventually led to the discovery of antimatter.

Doppler shift – The elongation or compression of a wavelength of light or sound, which depends on the motion of the emission source relative to an observer. For example, light's wavelength elongates as the galaxies farthest from Earth move away from us due to the expansion of space. The Doppler shift is toward the red portion of the spectrum, since red is the color of longer wavelengths of light.

Earth-like – Refers to a planet that is similar to Earth. Typically, the planet would be in the habitable zone of the star it orbits, be large enough for gravity to hold its atmosphere in place, and have liquid water, to name a few of the properties of an Earth-like planet.

Entropy – In nature, all processes produce waste energy (heat) that escapes and becomes unavailable to do mechanical work. The waste energy that escapes is a measure of the entropy, and increases the disorder of a system. Since the universe is continually evolving through cosmic processes, the entropy of the universe is always increasing.

Entropy apocalypse – Refers to a point in time when all energy is heat, and unavailable to do mechanical work. It spells doom for all life.

Euclidean geometry – The geometry developed by the ancient Greek mathematician Euclid. Mathematicians refer to it as a "flat" geometry, where parallel lines remain equidistant (the same distance apart) to infinity, and the sum of angles within a triangle equal 180 degrees. We typically learn this geometry as schoolchildren.

Existence Equation Conjecture – This theoretical equation enables the calculation of a mass' kinetic energy as it moves in the fourth dimension of Minkowski space. The equation is unusual in that it indicates the kinetic energy of a mass moving in the fourth dimension is negative, and massively large. A speculative interpretation of the Existence Equation Conjecture is that existence requires negative energy, which is being syphoned from the universe, and results in the accelerated expansion of the universe. The Existence Equation Conjecture is discussed more fully in Chapter 12, Appendix I and Appendix II.

Existential risk - Any risk with the potential to destroy humankind or drastically restrict human civilization. In theory, an existential risk could end the existence of Earth, the solar system, the galaxy, or even the universe.

Extraterrestrial life - Life that exists outside of the Earth. For example, if we discover microbial life on Mars, it would be proof of extraterrestrial life.

Fahrenheit - A temperature scale, at which water freezes at thirty-two degrees (32° F), and boils at two hundred twelve degrees (212° F).

Field - A concept in the physical sciences used to imply mediation at any distance, without the use of particles as mediators. For example, gravity exerts an attractive force between two masses via a gravitational field. If, however, science discovers the particle of gravity called the graviton, the mediation (the force-carrying particles) between the two masses will not be a field, but be by graviton particles. The field concept in physics is an old concept, and pre-dates understanding the role that particles play as mediators.

Fossil record - Generally refers to the total number of fossils found, and the information derived from them by paleontologists, which are scientist that study fossil (prehistoric) animals and plants.

Four fundamental forces – Modern physics recognizes four known fundamental forces, namely gravitational, electromagnetic, strong nuclear, and weak nuclear. All physical forces theoretically trace back to one or more of these forces.

Fourth dimension – In a 1912 manuscript on relativity, Einstein equated the fourth dimension to *ict* (where $i = \sqrt{-1}$, *c* is the speed of light in empty space, and *t* is time, representing the numerical order of physical events measured with "clocks"). The entire thrust of using four-dimensional space in the special theory of relativity is attributed to Russian mathematician Hermann Minkowski. In 1907, Minkowski demonstrated Einstein's special theory of relativity (1905), presented algebraically by Einstein, could be presented geometrically as a theory of four-dimensional space-time.

Galaxy – Refers to a system of millions to billions of stars, along with gas and dust, held together by gravity. Most stars, like our sun, have planets and other celestial bodies orbiting them. There are billions of galaxies in our universe. Typically, galaxies are separated from one another by vast regions of space (measured in light years).

Gamma rays – Refers to short wavelength, high-energy, electromagnetic radiation emitted by radioactive substances.

General theory of relativity – A theory developed by Albert Einstein dealing with gravity and non-inertial frames of reference. It is termed the "general theory of relativity" to differentiate it from the "special theory of relativity," which focused on inertial frames of reference.

Gravitational lensing – Refers to the phenomenon that light's path can be affected by gravity. For example, light from distant galaxies bends as it passes through the gravitational field of another galaxy. This can result in magnifying, distorting, or producing multiple images of the original light source for a distant observer.

Gravity (or Gravitation) – Refers to the attractive force one mass exerts on another mass.

Ground state – Refers to the lowest energy state of an atom or particle.

Ground-state entropy – Refers to the lowest entropy state of a system.

Heisenberg uncertainty principle – This scientific principle holds it is impossible to accurately determine both the position and velocity of a particle simultaneously. In the context of quantum mechanics, it conceptually conveys the probabilistic nature of physical phenomena at the atomic and subatomic level (quantum level).

Hubble volume – The region of the universe surrounding an observer, beyond which objects recede from the observer at a rate greater than the speed of light, due to the expansion of the universe.

Inertial frame of reference – A frame of reference at rest or moving at a constant velocity.

Kelvin – The International System of Units (SI) absolute thermodynamic temperature scale, using as its null point zero degrees Kelvin (0° K), which refers to a state devoid of all heat and motion. The Kelvin temperature scale equates to the Celsius temperature scale via the following equation: K = [°C] + 273.15.

Kinetic energy – The energy associated with an object due to its motion.

Lamb shift – This refers to the small difference in energy between two states of the hydrogen atom. American physicist Willis Eugene Lamb (1913-2008), first detected it, and received the Nobel Prize in Physics in 1955 for his discoveries related to the Lamb shift.

Macro-level – This is our everyday world. It is the reality that we can typically see and touch.

Mass – In physics, this typically refers to matter, such as a subatomic particle, atom, or an assembly of subatomic particles and atoms.

Mediators – In physics, mediators are the particles that carry force between entities. For example, the photon is the force carrier for the electromagnetic force.

Minkowski space – Refers to the mathematical, four-dimensional concept of space-time developed by Hermann Minkowski in his 1909 paper "Space and Time." Minkowski space-time found application in Einstein's special theory of relativity, and in the development of the Existence Equation Conjecture.

Miracle – This refers to an act by a supernatural being, which typically suspends the laws of physics.

Moore's Law – This is more of a general rule or observation than an actual physical law. Moore's law states the number of transistors that can be placed inexpensively on an integrated circuit doubles approximately every two years.

Multiverse – Refers to the concept that there are other universes beyond our own. The phrases, "parallel universes," "alternative universes," "quantum universes," "parallel dimensions," and "parallel worlds" are synonymous to the multiverse.

Muon – A negatively charged particle approximately 200 times more massive than an electron.

Neutrino – An elementary particle with close to zero mass, no electrical charge, and travels close to the speed of light.

Newton's laws of motion – See Classical Mechanics.

Newtonian mechanics – See Classical Mechanics.

Nomads – Refers to groups of people, typically a clan or tribe, whose civilization survives by moving from one location to another location, more favorable to their survival.

Occam's razor – A principle of science that holds the simplest explanation is the most plausible one, until new data to the contrary becomes available.

Particle – In physics, this can refer to a massless object or, alternately, a small object with mass. A photon is an example of a massless object. A muon is an example of a small object with mass.

Particle accelerator – Refers to apparatus that uses electromagnetic fields to accelerate subatomic particles to high velocities, even velocities approaching the speed of light.

Photoelectric effect – Refers to the ejection of electrons from any substance due to the incidence of electromagnetic energy (light).

Photon – Refers to a particle (energy packet) of light (electromagnetic radiation). The photon has no mass, and travels at the speed of light in a vacuum.

Planck length – The smallest unit of length theoretically possible, which suggests that space, itself, may be quantized. The Planck length is equal to approximately 1.6×10^{-35} meters, and is defined using three fundamental physical constants: the speed of light in a vacuum, Planck's constant, and the gravitational constant. At the Planck length, gravity is thought to become as strong as the three other fundamental forces (electromagnetic, strong and weak nuclear), and quantum effects dominate.

Planck time – This is the time it takes light in a vacuum to travel one Planck length. This is the smallest unit of time that science believes change can occur. It is approximately equal to 10^{-43} seconds. This implies that time itself may be quantized.

Products – Refers to the resulting substances of a chemical reaction. The substances may be compounds, elements, or both.

Quantized – Refers to the discrete nature of a substance, like mass or energy. In this book, it is typically used as a verb when describing the discrete nature of reality (mass, energy, space, and time).

Quanta – A discrete packet of energy, like a photon.

Quantum – Synonymous with quanta.

Quantum entanglement – Refers to a phenomenon in quantum mechanics where a pair of particles or photons interacts with each other, and forms an invisible bond. When a pair of particles becomes entangled, their quantum state, which completely describes their state of being, communicates and correlates with each other, even when the particles are separated by a distance. Thus, changing the quantum state of one entangled particle forces the quantum state of the other entangled particle to change in a way that they remain in a correlated quantum harmony. This phenomenon is a scientific fact, but not completely understood. One mystery is that the communication between entangled particles appears to travel faster than the speed of light in a vacuum.

Quantum fluctuation – An effect observed in quantum physics, where a temporary change in the amount of energy occurs at a point in space, in accordance with the Heisenberg uncertainty principle. This effect gives rise to "virtual particles" or "spontaneous creation." This is a scientific fact.

Quantum level – Refers to the scale of atoms and subatomic particles.

Quantum mechanics – Refers to the scientific principles that explain the behavior of physical objects and their interactions with energy on the scale of atoms and subatomic particles.

Quantum Universe – A theory that the entire universe consists of quantized matter and energy.

Quark – The quark is an elementary particle and a fundamental building block of other particles, like protons and neutrons. There are six types of quarks, known as flavors, including up, down, strange, charm, bottom, and top.

Qubit – In quantum computing, the qubit is the quantum bit of information analogous to the classical computer bit. Whereas the classical computer bit contains information, and can represent a 1 or 0, the qubit can represent a 1, 0, and a superposition of both at the same time. For example, using a classical computer bit, a polarized photon could be expressed by either a 1 for horizontal polarization or a 0 for vertical polarization. A cubit represents both states simultaneously.

Reactants – Refers to the substances (elements and compounds) that react in a chemical reaction to form one or more new substances (products).

Redshift – Refers to the elongation of a light wave, as its wavelength stretches, due to the emission source moving away from an observer. Longer wavelengths of light are in the red portion of the spectrum.

Relativistic mechanics – This refers to any form of mechanics that are derived from and/or compatible with Einstein's general and special theory of relativity.

Singularity – In the context of the Big Bang, this refers to the point of infinitely dense energy that gave birth to the Big Bang.

Space-time – This refers to the concept that time is dependent on space. In effect, space and time are fused together in a mathematical model to form a continuum known as space-time. This has enabled physicist to simplify numerous physical theories and describe the universe more precisely.

Special theory of relativity – A theory developed by Albert Einstein, based on two postulates:

> Physical laws have the same mathematical form in any inertial system (a system at rest or moving at a constant velocity).

The velocity of light is independent of the motion of its source, and will have the same value when measured by observers moving with constant velocity with respect to each other.

From these two fundamental postulates, Einstein was able to develop his famous mass-energy equivalence equation, the time-dilation equation, and the relativistic kinetic-energy equation. The special theory of relativity is one of the most successful theories of modern science.

Speed of light in a vacuum – This is the speed at which light (electromagnetic radiation) travels in a vacuum, which is exactly 299792458 meters/second. Scientists universally view light as the upper-speed limit in the universe. Nothing travels faster in a vacuum than the speed of light.

Spiral galaxy – A galaxy having a spiral form with spiral arms. The oldest starts in a spiral arm near the center of the galaxy. Our Milky Way galaxy is a spiral galaxy.

Spontaneous symmetry breaking – A theory that holds a system in a symmetrical state is able to transform to an asymmetrical state.

Standard Model – This refers to the Standard Model of particle physics, which mathematically models the behavior of elementary particles and their interaction relative to the electromagnetic, strong and weak forces.

Star – This is a self-luminous celestial body, like our sun, consisting of gas held together by gravity. The luminescence is the result of nuclear reactions within the body, whose energy makes its way to the surface, and emits as radiation.

String theory – A mathematical theory that represents all mater, such as subatomic particles, as consisting of strings that vibrate in one dimension, and exist in eleven dimensions. A number of prominent physicists consider string theory to be a contender for the theory of everything.

Supernatural being – A being that exists outside the natural realm. The word deity and god are synonymous with supernatural being. Depending on specific religious beliefs, the supernatural being has specific powers over the natural world.

Super-universe – See Bulk.

Symmetry of physical laws – A concept in physics that argues that a physical law is unchanged by any theoretical transformation. This is a simple example: A geometric sphere maintains all elements that define it as a geometric sphere, regardless of any rotational transformation. Most physicists believe physical laws are symmetrical.

Theory of everything – A self-contained mathematical model that describes all fundamental forces and forms of matter.

Thought experiment – This is a conceptual experiment. It considers a hypothesis, theory, or principle, and thinks through the ramifications, to illustrate a point. A thought experiment may or may not be possible to perform in reality. The objective is to explore the hypothesis, theory, or principle, and its potential consequences. Einstein is historically considered the master of the thought experiment, or "Gedankenexperiment" (in German).

Time – The traditional definition of time is the numerical sequence of events as measured by clocks.

Uncertainty principle – See the "Heisenberg uncertainty principle" above.

Velocity – The distance an object travels divided by the time it takes to travel that distance.

Virtual particle – A particle that exists for a limited time, and obeys some of the laws of real particles, including the Heisenberg uncertainty principle, and the conservation energy. However, their kinetic energy may be negative.

Wavefunction – The wavefunction, in quantum mechanics, describes the probability of a particle's state (position, momentum, and other attributes).

Wave-particle duality – Refers to the exhibition of both wave and particle properties.

Word line – The unique path that an object takes as it travels through four-dimensional space-time.

APPENDIX I

Derivation Existence Equation Conjecture

From the special theory of relativity, we know that the relativistic kinetic energy may be calculated using the following equation:

$$E_k = \frac{mc^2}{\sqrt{1-\frac{v^2}{c^2}}} - mc^2$$

Where E_k is the relativistic kinetic energy, m is the rest mass of an object, v is the velocity of an object, and c is the speed of light in a vacuum.

It is important to note that in the derivation of the relativistic kinetic energy, m is the rest mass of an object. The kinetic energy (E_k), derived using the object's rest mass and velocity, will result in the object becoming more massive via Einstein's mass-equivalence equation ($E=mc^2$). However, it is important to understand that m is the object's rest mass in the relativistic kinetic energy equation, since we will use the equation to derive the Existence Equation Conjecture.

From the discipline of calculus, we can calculate velocity along any coordinate by taking the first derivative of that coordinate with regard to time. Therefore:

$$v = \frac{dx}{dt}$$

Where x is a spatial coordinate, representing a unit of measure along the *x* axis.

In the special theory of relativity, Einstein used Minkowski's four-dimensional space: X_1, X_2, X_3, X_4, where X_1, X_2, X_3 are the typical coordinates of the three-dimensional space, and $X_4 = ict$, where $i = \sqrt{-1}$, *c* is the speed of light in empty space, and t is time, representing the numerical order of physical events measured with "clocks." (The mathematical expression i is termed an imaginary number because it is not possible to solve for the square root of a negative number.) Therefore, $X_4 = ict$, is a spatial coordinate, on equal footing with X_1, X_2, and X_3 (the typical coordinates of three-dimensional space). It is not a "temporal coordinate." This forms the basis for weaving space and time into space-time.

If we assume the object is at rest, X_1, X_2, and X_3 are each equal to zero, and the velocity associated with them is zero. However, the object continues to move in time along the X_4 coordinate. We can calculate the velocity associated with this movement by taking the first derivative of X_4 with regard to time. Therefore:

$$v = \frac{dX_4}{dt}$$

If we substitute $X_4 = ict$, the velocity becomes

$$v = ic$$

where $i = \sqrt{-1}$ and c is the speed of light in empty space

We can substitute the velocity into the relativistic kinetic energy equation as follows:

$$E_{kx4} = \frac{mc^2}{\sqrt{1-\frac{(ic)^2}{c^2}}} - mc^2$$

Substituting $i = \sqrt{-1}$ and simplifying yields the following results:

$$E_{kx4} = \frac{mc^2}{\sqrt{2}} - mc^2$$

$$E_{kx4} = \frac{mc^2}{\sqrt{2}} - mc^2$$

$$E_{kx4} = .707mc^2 - mc^2$$

$$E_{kx4} = -.3mc^2$$

Where m is the rest mass of the object, and c is the speed of light in a vacuum. Note, for simplicity, the equation is rounded to one significant digit after the decimal point.

The equation ($E_{kx4} = -.3mc^2$) is termed the Existence Equation Conjecture. To a first order, the Existence Equation Conjecture appears to agree with experimental time-dilation results (see Appendix II). Until we have more data, and the scientific community weighs in on its validity, I have labeled the equation a conjecture, hence its name: Existence Equation Conjecture.

APPENDIX 2

Experimental Verification Existence Equation Conjecture

From particle accelerator data (Bailey, J. et al., Nature 268, 301 (1977) on muon lifetimes and time dilation.): In an experiment at CERN by Bailey et al., muons of velocity 0.9994c were found to have a lifetime 29.3 times the laboratory lifetime.

If a muon needs $KE_{X4} = -.3mc^2$ at rest for one lifetime, per the Existence Equation Conjecture, conceptually it requires the following negative energy to exist for 29.3 lifetimes:

$$E_{X4} = 29.3 \text{ x } -.3mc^2$$

(In effect, this argues the amount of negative energy required for existence increases. This suggests that an enormous amount of positive energy would be needed to cancel the negative energy of existence.)

If we do the mathematics, we get the following result:

$$E_{X4} = -8.8mc^2$$

We must judge that the kinetic energy is completely consumed when the particle ceases to exist (decays) as a muon. (We know that muons decay to one electron and two neutrinos, and sometimes the decay produces other particles, like photons.) Please note, in keeping with the accuracy of the lifetime data (observed time dilation), we will round off all calculations of kinetic energy to one decimal point.

We will consider the Bailey experiment, and calculate the relativistic kinetic energy of a muon traveling at .9994*c*:

Relativistic calculation of the muon's kinetic energy from Einstein's relativistic kinetic energy equation, delineated below:

$$E_k = \frac{mc^2}{\sqrt{1 - \frac{v^2}{c^2}}} - mc^2$$

If we let $v = .9994c$, $v^2 = .9988c$, reducing to two significant digits (not rounding up to 1.0*c* to avoid conflict with the theory and an irrational result) implies $v^2 \sim .99c^2$

Then,

$$E_k = \frac{mc^2}{\sqrt{.01}} - mc^2$$

$$E_k = \frac{mc^2}{.10} - mc^2$$

$$E_k = 10\,mc^2 - mc^2$$

$$E_k = 9.0mc^2$$

Therefore, the kinetic energy of a muon accelerated to approximately 99.9% the speed of light is approximately $9.0mc^2$

Discussion of Results:

Theoretically, the Existence Equation Conjecture predicts – $8.8\ mc^2$ of negative energy is required for a 29.3 lifetime extension (the time dilation we observe).

The experimental kinetic energy (one decimal point accuracy) is calculated to be $9.0\ mc^2$. This positive kinetic energy completely balances the negative energy within ~ 2%. Given the experimental accuracy and the rounding errors, this is remarkably close agreement.

What are the decay products of a muon at rest? To address this question, we must postulate the following:

1) The energy associated with the rest mass of the muon is $E = mc^2$ (Einstein's mass-energy equivalence equation).

2) The energy for the lifetime (~2.2 x 10^{-6} seconds) of the muon to exist at rest is $KE_{X4} = -.3mc^2$ (Existence Conjecture Equation).

3) All muons decay to one electron and two neutrinos. Sometimes the decay produces other particles, like photons.

Based on the above, it can be inferred the rest mass energy ($E = mc^2$) is consumed as follows: $.3mc^2$ is required to satisfy the Existence Conjecture Equation ($E = -.3mc^2$), which gives the muon its at-rest lifetime. The remaining $.7mc^2$ produces the decay particles (electron, two neutrinos, and potentially other particles like photons). Although the lifetimes of muons at rest have been accurately measured, the energy of the decay products has not. Therefore, further data is required to validate this inference.

APPENDIX III

Time Dilation Experimental Evidence Introduction

Please be advised that the items below represent a sample from the vast amount of experimental evidence available. Their selection rests on historical relevance and scientific validity.

Velocity Time-Dilation Experimental Evidence

Rossi and Hall (1941) compared the population of cosmic-ray-produced muons (an elementary particle similar to and more massive than an electron) at the top of a mountain to muons observed at sea level. A muon is a subatomic particle with a negative charge. Muons occur naturally when cosmic rays (energetic-charged subatomic particles, like protons, originating in outer space) interact with the atmosphere. Muons, at rest, disintegrate in about 2 x 10^{-6} seconds. The mountain chosen by Rossi and Hall was high. The muons should have mostly disintegrated before they reached the ground. Therefore, extremely few muons should have been detected at ground level, versus the top of the mountain. However, their experimental results indicated the muon sample at the base experienced only a moderate reduction. They made use of Einstein's time-dilation effect to explain this discrepancy. They attributed the muon's high speed to be dilating time. In this experiment, the muons were decaying approximately 10 times slower than if they were at rest.

In 1963, Frisch and Smith, once again confirm the Rossi and Hall experiment, proving beyond doubt that extremely high kinetic energy prolongs a particle's life.

With the advent of particle accelerators, capable of accelerating particles to nearly the speed of light, the confirmation of the time-dilation effect is routine. There are far too many accelerator experiments to list them all, but one in particular I think merits our attention. It confirmed the "twin paradox." The twin paradox goes something like this. A twin makes a journey into space in a near speed-of-light spaceship, and returns home to find he has aged less than his identical twin that stayed on Earth. This means that clocks sent away at near the speed of light, and returned near the speed of light to their initial position, demonstrate retardation (record less time) with respect to a resting clock.

The clock at rest will show a greater elapse in time. The clock that journeyed at near the speed of light will show a smaller time elapse. This is a proven scientific phenomenon.

In 1977, Bailey and his colleagues measured the lifetime of muons accelerated in the European Organization for Nuclear Research (CERN) muon storage ring. This circular ring enables particles to accelerate at a high velocity over a prolonged period. As they sent the particles around the loop, they measured time dilation, and found the life of the particle extended. This means that when the particle returned to its initial starting position, time remained dilated. Therefore, this experiment is widely held to confirm the twin paradox.

Gravitational Time Dilation Experimental Evidence:

In 1959, Pound and Rebka measured a slight redshift in the frequency of light emitted close to the Earth's surface (where Earth's gravitational field is higher), versus the frequency of light emitted at a distance farther from the Earth's surface. The results they measured were within 10% of those predicted by the gravitational time dilation of general relativity.

In 1964, Pound and Snider performed a similar experiment and their measurements were within 1% predicted by general relativity.

In 1980, the team of Vessot, Levine, Mattison, Blomberg, Hoffman, Nystrom, Farrel, Decher, Eby, Baugher, Watts, Teuber, and Wills published, "Test of Relativistic Gravitation with a Space-Borne Hydrogen Maser," and increased the accuracy of measurement to about 0.01%. In 2010, Chou, Hume, Rosenband, and Wineland published, "Optical Clocks and Relativity." This experiment confirmed gravitational time dilation at a height difference of one meter, using optical atomic clocks.

Made in the USA
San Bernardino, CA
08 August 2013